T0295122

POLYMER SCIENCE AND TECHNOLOGY

POLYMER RESEARCH AND APPLICATIONS

POLYMER SCIENCE AND TECHNOLOGY

Additional books in this series can be found on Nova's website under the Series tab.

Additional E-books in this series can be found on Nova's website under the E-books tab.

POLYMER SCIENCE AND TECHNOLOGY

POLYMER RESEARCH AND APPLICATIONS

ANDREW J. FUSCO
AND
HENRY W. LEWIS
EDITORS

Nova Science Publishers, Inc.
New York

LIBRARY OF CONGRESS CATALOGING-IN-PUBLICATION DATA

Polymers research and applications / editors, Andrew J. Fusco and Henry W. Lewis.
p. cm.
Includes index.
ISBN 978-1-61209-029-0 (hardcover)
1. Polymers. I. Fusco, Andrew J. II. Lewis, Henry W.
TA455.P58P6958 2010
620.1'92--dc22
2010047081

Published by Nova Science Publishers, Inc. ✝ New York

CONTENTS

PREFACE

This book presents and discusses research in the study of polymers. Topics discussed include extrusion heads used in manufacturing polymeric castings; effects of polypropylene fibers on expansion properties of cement-based composite containing waste glass; biodegradation of film polymer coating; new polymer technologies with water and physical-chemical properties of natural polymers.

Chapter 1 – On the basis of executed theoretical and experimental investigations highly productive constructions of extrusion heads used in manufacturing polymeric casings for preventing pipes from corrosion were worked out. Constructions of extrusion heads used in manufacturing of outer and inner casing for pipes are represented in the following paragraphs. The order of calculations of forming channels of extrusion heads are represented here too.

Chapter 2 – The amount of waste glass sent to landfill has increased over recent years due to an ever growing use of glass products. Landfilling can cause major environmental problems because the glass is not biodegradable material. However, waste glass can be used as fine aggregate, coarse aggregate and powder form in concrete. The fine and coarse aggregate can cause alkali-silica reaction (ASR) in concrete, but the powder form can suppress their ASR tendency and acts as a pozzolanic material. This paper studies the size effect of waste glass powder as a partial replacement in cement and its pozzolanic behavior. Silica fume and rice husk ash also used as a replacement part of cement for their property comparison with waste glass.

Also, ASR expansion was different volume fraction of polypropylene fiber in anticipation of reducing ASR expansion.

Results show that 10% replacement of Portland cement weight by finely waste ground glass (under 75μm), silica fume and rice husk ash can decrease ASR expansion sharply. Moreover, utilization of PP fiber more then 0.5 Vol% can suppress the excessive expansion in the matrix.

Finally combination of high reactive pozzolans and PP fiber is the best way to eliminate the harmful effects of ASR phenomena in the world.

Chapter 3 – The present paper deals with enzymatic biodegradation of film chitosan coatings which can be used for protecting open wounds (burning, surgical ones) and the means of their modification for extending the service life.

Chapter 4 – The description of the construction, the principle of action and calculation method of geometric parameters of extrusion head for granulating to reduced to fragments polymeric waste materials are stated in this work. Experimental data of the process of granulating are described here too.

Chapter 5 – The present research shows the high aggressiveness on the polymeric material and articles and it specially shows the aggressive factors of the humid and tropical climate of Cuba on the rubbers the different and more significative atmospheric factors, that is to say (temperature, solar radiation, ozone concentration) were supplied by the Instituto de Meteorología de la República de Cuba (The Institute of Meteorology of the Republic of Cuba). The samples of the tested rubbers, especially elaborated with no antioxidant protections were made using the coded rubber, used by the Rubber enterprise Union with the number 471.

The evaluation of the spouting and development of the ageing of the rubber samples mounted on the resting frames and under tension was carried out according to prior studies based on internationals standards.

The early apparition of cracks (days of exposition perpendicular to the direction they were mounted on the testing frames is a demonstration of the high aggressiveness of the humid and tropical climate of Cuba.

The results obtained and presently shown in this work ,show, on one side ,to give a good antioxidant protection to the rubber and their articles to be used under the atmospheric condition under the humid, tropical climate of Cuba and also that the atmospheric testing stations used could be employed to select antioxidant protection on systems of high protective effectiveness

Chapter 6 – Cross-linking polyethylene PEX-a were investigated. Cross-linking polyethylene pipes are exposed for boiling in hot water at 95oC. Analysis of migration antioxidants into hot water were investigated by Fourier-IR-spectroscopy method, UV-spectroscopy and liquid chromatography method (HPLC). Heat stability formulations PEX-a were investigated by DSC (method oxidation induction time OIT).

Chapter 7 – Thermodynamic characteristics and physical-chemical properties of natural polymers (cellulose, starch, agar, chitin, pectin and inulin), their water mixtures and some biologically active substances extracted from vegetable substances using carbon dioxide in a supercritical state are reviewed. In addition, several aspects of practical application of thermodynamic characteristics of biologically active substances are demonstrated.

Chapter 8 – A number of metal-containing polyamines have been synthesized employing the condensation process, generally one of the interfacial polycondensation processes. The majority of these have been synthesized from the reaction between mono and diamines and Group IVB metallocene dihalides and organotin dihalides. The major reasons for the synthesis of the Group IVB polyamines is to control light and as control release agents and more recently for electrical applications. Ruthenium-containing polyamines were synthesized as part of a solar energy conversion effort. The major reasons for the synthesis of the organotin polyamines is biological since many of these materials offer good ability to resist a wide variety of bacteria, cancer cell lines, and viruses.

Chapter 9 – 2-diethylaminoethyl(DEAE)-Dextran is still an important substance for transfection of nucleic acids into cultured mammalian cells by the reason of its safety owing to autoclave sterilization different from lipofection vectors. However, DEAE-Dextran may not be superior to lipofection vectors with cytotoxic and a transfection efficiency. A stable soap-less latex of 2-diethylaminoethyl(DEAE)-Dextran-methyl methacrylate(MMA) graft copolymer (DDMC) of a high transfection activity has been developed as Non-viral gene delivery vectors possible to autoclave at 121 for 15 minutes.Transfection activity determined by the X-gal staining method show a higher value of 50 times or more for DDMC samples than for the starting DEAE-dextran hydrochloride and a low cytotoxic is observed for DDMC. DDMC has been also observed to have a high protection facility for DNase degradation.. The resulted DDMC on grafted MMA, to form a polymer micelle of Core-Shell particle, should become a stable latex with a hydrophilic-hydrophobic micro-separated-domain. The complex by DDMC/DNA may be formed initially on the stable spherical structure of the amphiphilic micro-separated-domain of DDMC and have a good affinity to cell membrane for the endocytosis. The stronger infrared absorption spectrum shift to a high energy direction at around 3450cm−1 of the complexes between DNA and DDMC compared with DEAE-Dextran may mean to form more compact structures not only by a coulomb force between the phosphoric acid of DNA and the diethyl-amino-ethyl(DEAE) group of DEAE-Dextran copolymer but also by a force from multi-inter-molecule hydrogen bond. It should conclude to DNA condensation by these inter-molecular multi-forces to be possible the higher transfection efficiency. The complex should be formed following Michaelis-Menten type equation such as complex = K1(DNA)(DDMC). It should be supported that DDMC has a strong adsorbing power with DNA because of not only its cationic property but also its hydrophobic bond and hydrogen bond. The high efficacy of this graft-copolymer autoclave-sterilized for transfection can make it a valuable tool for a safety gene delivery.

Commentary – Limited resources of water, energy, and materials cause a growing need for new concepts in material development and its ecological integration. Water technology and resource-management is a central factor not only in environmental protection, in health care, and agriculture, but also in energy production and for the availability of biomass as a material source. Within Polymer Science, this requires an adoption of a ‘biological view’ on polymer synthesis, application, and the fate of polymer materials after use. Here, a key to future advancement is the ability to deal with water and to take advantage of its unique properties. Due to its dielectric properties, amphoteric nature and ability to undergo hydrogen bonding, water is certainly the most powerful solvent the authors know. It imposes enthalpic and entropic forces on other molecules that form an essential base for the richness in structure formation found in nature and the molecular functions in living systems. Hence, very topical scientific and technological challenges open up upon mastering water-based self-organisation of polymers; responsiveness, switching water solubility for water insolubility, and hybridisation of synthetic components with biological systems.

Versions of these chapters were also published in *Polymers Research Journal,* Volume 3, Numbers 1-4, published by Nova Science Publishers, Inc. They were submitted for appropriate modifications in an effort to encourage wider dissemination of research.

In: Polymer Research and Applications
Editors: Andrew J. Fusco and Henry W. Lewis

ISBN: 978-1-61209-029-0

Chapter 1

Extrusion Heads Used in Manufacturing Polymeric Casings for Preventing Pipes from Corrosion

A. K. Panov, T. G. Beloborodova, T. A. Anasova and G. E. Zaikov
Academy of Science of Republic of Bashkortostan,
Sterlitamak Branch, Sterlitamak, Russia
N.M. Emanuel Institute of Biochemical Physics, RAN, Moscow, Russia

Abstract

On the basis of executed theoretical and experimental investigations highly productive constructions of extrusion heads used in manufacturing polymeric casings for preventing pipes from corrosion were worked out. Constructions of extrusion heads used in manufacturing of outer and inner casing for pipes are represented in the following paragraphs. The order of calculations of forming channels of extrusion heads are represented here too.

Keywords. Extrusion Heads, polymer materials, manufacturing, polymeric casings.

1. Construction of Extrusion Heads Used in Manufacturing Casing for Outer Pipe Framing

Polymeric materials due to their qualities find their more and more usage as construction materials, so allowing producing different articles including polymeric casings for pipes. Polymeric casing looks like a hollow body with a cylindrical outer surface and there are longitudinal lugs in the inner surface. Longitudinal lugs serve as a support for a pipe which is inside the casing. Thus, according to Figure 1, polymeric casing (1) frames pipe (2) and because of longitudinal lugs make channels (4) between outer surface of pipe and inner surface of casing (1).

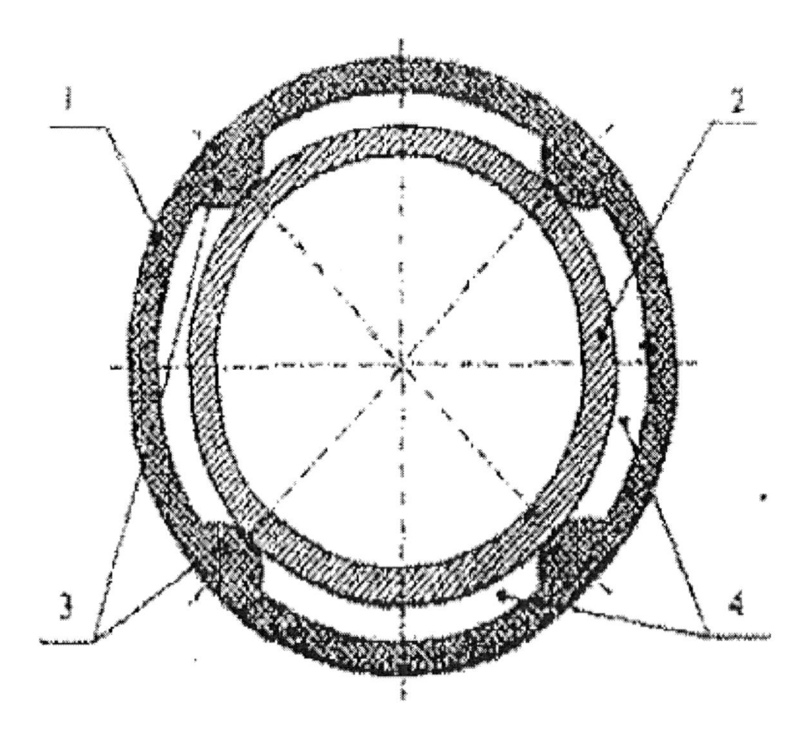

Figure 1. Pipe cross-section in casing.

Mentioned above construction of casing allows not only to isolate pipe from environment damages, saving it from corrosion and lowering heat exchange through the walls, but it also prevents overflow of transported product when hermetidy of pipe is damaged. And thus environment soiling is excluded.

We offer to use extrusion head used for manufacturing polymeric casing which is comprised of disconnected body, parts of which are tied by means of connecting bolts; mandrel-keeper, cone, grating matrix and disconnected mandrel, which form shaping channel. At the same time mandrel is executed with longitude slots, surfaces of which are formed by a curve of the second order and with a smooth section in the place of connecting of longitude slot with a cylindrical part of mandrel surface. There is an opportunity of cooling of inner surface of receiving casing [1, 2].

Extrusion head used for manufacturing polymeric casings, according to figures 2 and 3, is comprised of body (1),inside which cone (3) is located with the help of mandrel-keeper (3). In mandrel-keeper there are some slots, in which there is a pipe for water outlet (4) and a pipe for water supply (5). Fixing of mandrel-keeper (3) in the head is carried out with the help of the part of body (6), where regulating bolts (7) are established. They serve for radial removing of matrix (8). Longitude slots (10) are in mandrel (9) for shaping support lugs in polymeric casing, and at the same time in the place of connecting longitude slot (10) with cylindrical part of the surface of mandrel (9), a smooth section is made along the arch of a circle. Heating of extrusion head is made by electric heaters (11). The inner surface of cavity for melt current form parts of mandrel (12) and (13). In the cavity for melt current there is a cylindrical grating (14). A pipe for water feeding into calibrating mechanism (15) is placed inside of bar (16). Bar (16) is fixed in the base of bar (17), which is heat isolated with the help heat steady packing (18).

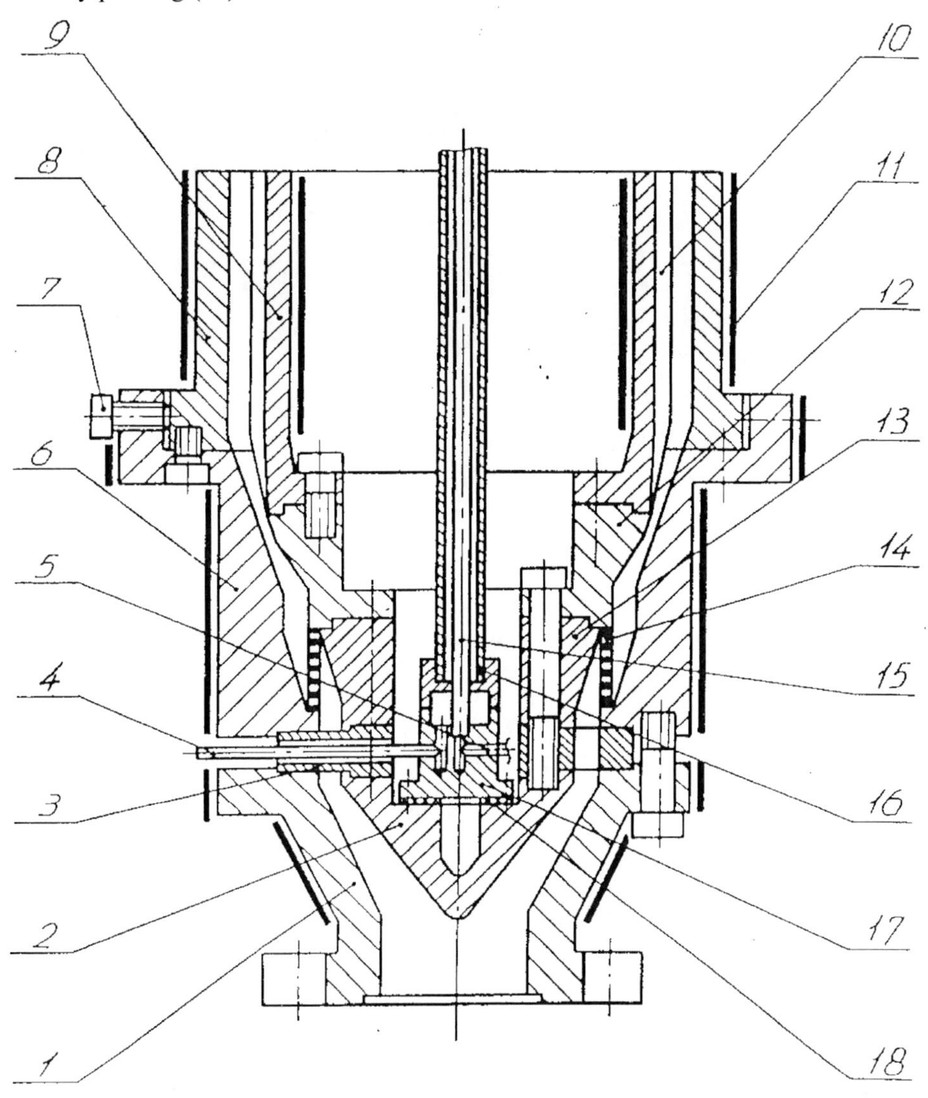

Figure 2. Longitude section of extrusion head used for manufacturing outer polymeric casings.

Shaping channel is formed by matrix (8) and mandrel (9). Cross section of shaping channel repeats the shape of cross section of polymeric casing pipe framing. Surfaces of longitude slots (10), made in mandrel (9) are formed by the second order curve, for example by a curve $Y=kx^2$, where k is the coefficient of pressing of parabola branches, mm^{-1}, which is chosen according to the equation condition of pressure overfall in the ring part and in the slot part of shaping channel (e.g. for extrusion head, shaping channel length – 150mm, outer diameter of shaping channel – 96 mm; inner diameter of shaping channel – 76 mm and longitude slot depth – 8 mm; coefficient $k=0{,}024mm^{-1}$). Longitude slots (10) form support lugs in the inner surface of polymeric casing, providing equation of pressure overfall along all the section of shaping channel of extrusion head.

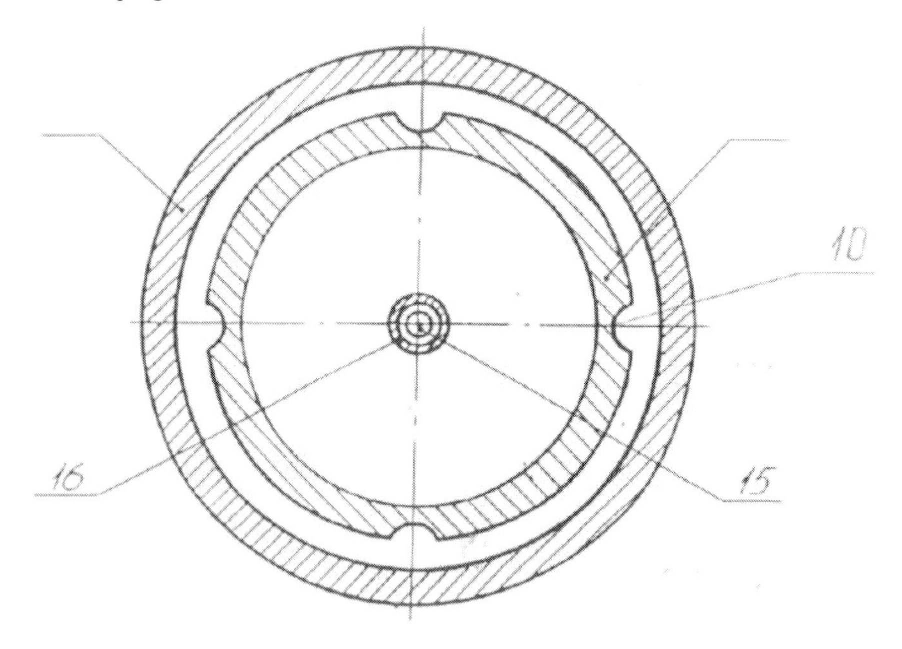

Figure 3. Cross-section of extrusion head used for manufacturing polymeric casings of outer pipe framing.

2. Description of Construction Extrusion Head Used for Manufacturing Polymeric Casings of Inner Pipe Framing

The developed design of extrusion head used for manufacturing polymeric casings of inner pipe framing provides hightened strength characteristics, increases hardness of polymeric casing of inner pipe framing because of support lugs on its outer surface are shaped. Increasing of quality and dimensional precision of polymeric casing of inner pipe framing is achieved thanks to the fact that making longitudinal slots in the inner surface of matrix makes it easier to calibrate casing. Disposition of longitudinal slots on the inner surface of matrix allows to shape support lugs on the outer surface of polymeric casing of inner pipe framing and this increases hardness of polymeric casing and also its strength characteristics.

Such a design makes the process of calibrating of polymeric casing because it is done on the outer diameter with the help of calibrating nozzle, connected with cooling bath. As a result it is not necessary to use parts, providing inner cooling of casing and the quality of outer of outer surface of polymeric casing increases.

Figure 4 demonstrates longitudinal section of extrusion head used for manufacturing polymeric casing of pipe inner framing. Figure 2 demonstrates a cross-section of a moulding channel of extrusion head used for manufacturing polymeric casing of pipe inner framing.

According to Figure 4 an extrusion head used for manufacturing polymeric casing of pipe inner framing consists of case (1) within which cone (2) is set by means of mandrel-keeper

(3). Fixation of mandrel-keeper (3) in the head is provided by means of case (4) detail, within which regulating bolts (5) for a radial displacement of matrix (6) are set.

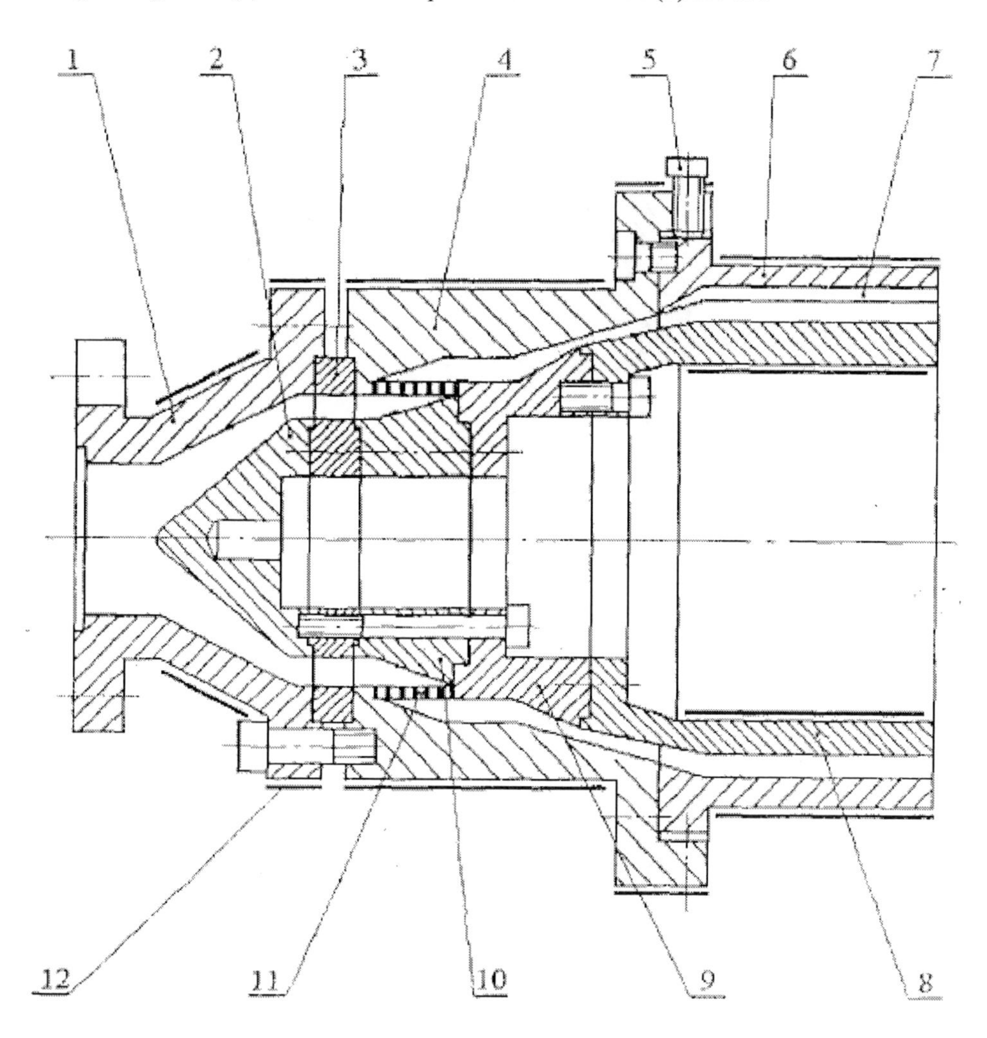

Figure 4. Extrusion head for manufacturing polymeric casing of pipe inner framing.

To form support lugs in extruded polymeric casing in matrix (6), longitudinal slots are made; simultaneously there is a smooth section of matrix along the circumference's arch in a part where a longitudinal slot surface joins a cylindrical part of matrix surface. The hollow inner surface for polymers melt flowing is formed by mandrel (8) and its details (9) and (10). For melt flowing there is a cylindrical grating (11) in a hollow. Heating of extrusion head is provided by electrical heaters (12).

Extrusion head used for manufacturing polymeric casing of pipe inner framing works as follows. Having gone through the filter polymer's melt is fed to a round cross channel situated in case (1), according to Figure 1 by means of cone 2 polymer's stream is cut and gets the ring shape. While passing through mandrel-keeper the melt is cut by mandrel-keeper's ribs. To provide welding ability of polymer's stream there is a cylindrical grating in the shape of a glass, set behind mandrel-keeper.

The hollow of feeding channel is farmed by a case detail (4) and mandrel's details (9, 10). Just in front of mandrel (8) melt stream passes the section of narrow spot, this allows to improve the welding ability of polymer's melt stream.

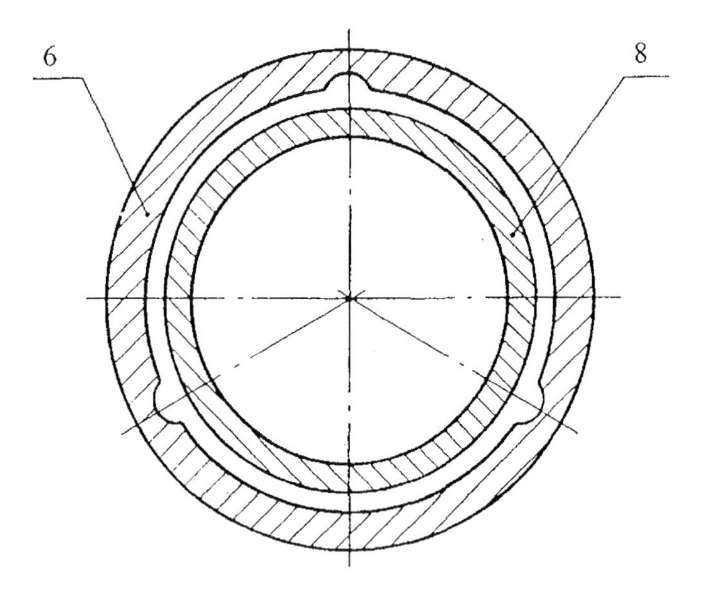

Figure 5. Cross-section of extrusion head for manufacturing pipe inner framing.

According to Figure 5 a moulding channel is formed by matrix (6) and mandrel (8). The shape of a moulding channel cross-section coincides with a shape of cross-section of polymeric casing of pipe inner framing.

The longitudinal surfaces (7) in matrix (6) are formed by the second order curve. The longitudinal slots form support lugs on the outer surface of polymeric casing of pipe inner framing; providing equality of pressure overfall all section of extrusion head moulding channel long. A smooth matrix cutoff along the circumference's arch in a place where longitudinal slots' surfaces join a cylindrical part of a matrix surface eliminates cutting off a profile of polymeric casing of pipe inner framing diminishing shear strength in a polymer melt stream in longitudinal slots.

Extrusion head has outer and inner heating, provided by means of electrical heaters (12). Openings for thermo pairs setting within them are provided for head details, their number depends on the number of independently regulated parts of heating. The outer surface framing casing is formed by matrix (8), which position is changeable depending on its displacement to radial direction by means of regulating bolts (7). To supply a calibrating device with cooling water and to draw it aside the following details are used: a tube for water pipe-bend (4), supply tube (5), a tube for water supply into calibrating device (15), bar (16) and bar base (17). By means of these details before hand cooling of polymeric casing inner surface becomes possible.

An extrusion head is heated by electrical heaters (11). Openings for setting thermo pairs within them are provided for head details in accordance with the number of independently regulated heating parts. Inner extrusion head filled with air and a thermo stable pad (18) as well, essentially reduce an unwilling heat exchange between a hot head and the details providing cooling water supply and pipe-bend.

Thus, the design of extrusion head for manufacturing outer polymeric casing is analogous to the design of extrusion head for manufacturing polymeric pipes. The difference is in the structure of shaping channel, i.e. the cross-section of polymeric casing (Figure 4). Correspondingly, in a shaping channel there is a section of longitudinal slots parallel to ring's section (like in extrusion heads for pipe manufacturing), which forms longitudinal lugs on the outer casing surface.

3. Description of Extrusion Head of Calculation Method Used for Manufacturing Polymeric Casing of Pipe Framing

Designing the extrusion heads construction required selection of polymeric material for casing manufacturing as well as calculation of its dimensions taking into account the outer diameter of steel pipe-line. Polymeric casing is designed for its framing.

Having taken into consideration the most frequently used polymeric materials, such as high density polyethylene (HDP), low density polyethylene (LDP), polypropylene (LP), polyvinylchloride (PVC) as material for manufacturing polymeric casing of pipe framing, polyvinylchloride was chosen. The selection is explained by PVS's advantages to other large capacity plastics. They are: the raw material availability, the durable qualities, taking care of natural resources, solving ecological problems.

In our work we use PVC composition with mixing of the following components (%): PVC- 77,0-91,0; calcium stearat – 2,5; plumber silicate – 1,2; dioetilphtalate – 3,0-20,0.

To ascertain polymeric casing dimensions, thickness of its wall S and height of support lug h [3] were calculated. The calculation of height h was carried out on permissible contact stress $[\sigma_k]$ which is equal to 3-8 MPa for a selected polymeric material.

It was supposed that a cross-section of support lug is semicircle with radius h and that steel pipe with radius R is based on two support lugs.

A final formula for calculating the height of support lug is as follows:

$$h = \left(\frac{[\sigma_k]^2 \cdot \pi \cdot (1-\mu^2) \cdot (E_1 + E_2)}{g_E \cdot E_1 \cdot E_2} - \frac{1}{R} \right)^{-1}, \tag{1}$$

where E1, E2 - are elasticing modules of casing and pipe-line materials;
qk –force given for a contact length unit.

The calculation of wall thickness was carried out under the following conditions: stability of polymeric casing shape while pipe-line laying in the ground, durability of casing material under pressure which is equal to a working one in a pipe-line being framed.

Carrying out the calculation of polymeric casing concerning durability of shape, it was supposed that outer framing surface is under inner pressure P, which is equal to the pressure of ground on the casing surface:

$$P = G_Г / S_{ОБ}, \tag{2}$$

where Gr - gravity force of the ground over the casing;

SoБ – casing surface influenced by gravity force G_R.

Critical outer pressure of casing pkp was calculated with a formula:

$$p_{кр} = \frac{E \cdot S^3}{4 \cdot (1 - \mu^2) \cdot R^3}, \tag{3}$$

where E is elasticity module of casing material; R is inner radius of a cylindrical part of casing.

It is possible to calculate the minimum thickness of casing wall S for condition of shape stability taking into account the condition p ≤ pkp and introducing the coefficient of stability reserve:

$$S = R\sqrt{\frac{4m\rho(1-\mu^2)}{E}}. \tag{4}$$

To provide casing material durability under working P in a framed pipe-line, the pipe framing casing is to bear a working pressure of substance pumped over the framed pipe-line. So the calculation of wall thickness is carried out proceeding from permissible strain for a selected material σ_g, at the same time taking into account the maximum pressure P for a selected material.

Thus, we conclude that the suggested construction of an extrusion head for manufacturing polymeric casing allows producing pipe-line casings with dimensions and quality required.

REFERENCES

[1] Patent RF № 2134640, 6B29C47/20 *Extrusion head used for manufacturing polymeric casing pipe framing* / A.K. Panov, A.N. Mehlis(RF). Applicated 24.02.98. Published 20.08.99.

[2] A.K. Panov, A.N. Mehlis, A.A. Panov, *Elaboration of extrusion head construction used for manufacturing polymeric casing* // Technique on the bourderline of the 21 century: Collection of scientific works. Ufa: Gigem. 1999. p. 48-57.

[3] A.K. Panov, A.N. Mehlis. *Method of dimension calculation of polymeric casing pipe framing* // Cybernetics method of chemical-technological processing: Collection of reports at scientific conference. Ufa: USOTU. 1999. V.2. Book 1, p.115.

In: Polymer Research and Applications
Editors: Andrew J. Fusco and Henry W. Lewis

ISBN: 978-1-61209-029-0

Chapter 2

EFFECTS OF POLYPROPYLENE FIBERS ON EXPANSION PROPERTIES OF CEMENT BASED COMPOSITE CONTAINING WASTE GLASS

A. Sadrmomtazi, O. Alidoust and A. K. Haghi*
University of Guilan, Rasht, Iran

ABSTRACT

The amount of waste glass sent to landfill has increased over recent years due to an ever growing use of glass products. Landfilling can cause major environmental problems because the glass is not biodegradable material. However, waste glass can be used as fine aggregate, coarse aggregate and powder form in concrete. The fine and coarse aggregate can cause alkali-silica reaction (ASR) in concrete, but the powder form can suppress their ASR tendency and acts as a pozzolanic material. This paper studies the size effect of waste glass powder as a partial replacement in cement and its pozzolanic behavior. Silica fume and rice husk ash also used as a replacement part of cement for their property comparison with waste glass.

Also, ASR expansion was different volume fraction of polypropylene fiber in anticipation of reducing ASR expansion.

Results show that 10% replacement of Portland cement weight by finely waste ground glass (under 75μm), silica fume and rice husk ash can decrease ASR expansion sharply. Moreover, utilization of PP fiber more then 0.5 Vol% can suppress the excessive expansion in the matrix.

Finally combination of high reactive pozzolans and PP fiber is the best way to eliminate the harmful effects of ASR phenomena in the world.

Keywords: Waste glass, pozzolanic behavior, particle size effect, Expansion, Fiber reinforcement.

* Corresponding author e-mail: Haghi@Guilan.ac.ir

1. INTRODUCTION

The amount of waste glass has gradually increased over recent years due to an ever-growing use of glass products. Most colorless waste glasses have been recycled effectively. On the other hand, colored waste glasses, with their low recycling rate, have been dumped into landfill sites. However, with a shortage of landfill sites, landfilling them is becoming more and more difficult. Additionally, landfilling of waste glasses is undesirable because they are not biodegradable, which makes them environmentally less friendly. Considering these facts, the reutilization of colored waste glasses has drawn more attention in recent years.

Current reprocessing costs of waste glasses for special use are rather high. However, when waste glasses are reused in making concrete, the production cost of waste glasses for concrete will go down through the development of reprocessing technology and the extension of reprocessing facilities, which will make concrete containing waste glasses economically viable.

Studies have been done on the possibility of reusing waste glasses as asphalt additive or road filler [1 2]. Waste glasses were used as aggregates for concrete [3–5]. However, the applications were limited due to the damaging expansion in the concrete caused by alkali–silica reaction (ASR) between high-alkali pore water in cement paste and reactive silica in the waste glasses. The chemical reaction between the alkali in Portland cement and the silica in aggregates forms silica gel that not only causes cracks upon expansion, but also weakens the concrete and shortens its life [6]. Recently, studies have been carried out to suppress the ASR expansion in concrete and find methods to recycle waste glasses [7,8].

Polypropylene fibers, produced by the fibrillation of polypropylene films, have been used in Portland cement concrete since the late 1960s [9]. Polypropylene and other synthetic fibers are added to concrete as secondary reinforcement to control plastic shrinkage w x [10]. The most common application is slab-on-grade construction where the constraint of the foundation or other parts of the structure produces tensile stresses when the concrete shrinks due to moisture loss. These stresses may exceed the concrete strength at early age leading to shrinkage cracks. Polypropylene fibers mitigate plastic and early drying shrinkage by increasing the tensile concrete and bridging the forming cracks. The effect of polypropylene fibers on the properties of hardened concrete varies depending on the type, length, and volume fraction of fiber, the mixture design, and the nature of the concrete materials used.

It also revealed the expansion and strength characteristics of concrete due to ASR by mixing reinforcing polypropylene fibers [11] to suppress the expansion by ASR when using waste glasses as aggregates.

In this research project, some mixtures were prepared by composite cement was made by replacement of %10 of weight of cement by pozzolans and different volume fracture of PP fiber in the matrix for studying the effects of this materials to control the amount of expansion due to alkali-silica reaction.

2. MATERIAL SPECIFICATIONS

All mixtures that contained different fiber volume fractions percentages and silica fume, rice husk ash and glass as partial replacement of cement were fabricated and tested in order to assess fresh and hardened properties of FRSFEC. Materials, specimen fabrication, curing conditions, and testing methods used in this investigation were designed to stimulate potential repair applications of FRSFEC in practice.

Materials used included Type K expansive cement, silica fume, rice husk ash, glass, coarse aggregate with a maximum nominal size aggregate of 16 mm, concrete sand, tap water, high-range water-reducing admixture HRWRA and fibrillated polypropylene fibers. The fiber lengths used ranged 6 mm that prepared and cut in the factory. Clean flat glass was used in this study. The chemical composition of the glass and other materials was analyzed using an X-ray microprobe analyzer and listed in Table 1 together with that of silica fume and rice husk ash for comparison.

Table 1. Chemical composition of glass, silica fume and rice husk ash (by weight percent)

Materials	Sio_2	Al_2O_3	Fe_2O_3	CaO	MgO	Na_2O	K_2O	CL	SO_3	$L.O.I$
Glass	72.50	1.06	0.36	8.00	4.18	13.1	0.26	0.05	0.18	-
Silica fume	91.1	1.55	2.00	2.24	0.60	-	-	-	0.45	2.10
Rice husk ash	92.15	0.41	0.21	0.41	0.45	0.08	2.31	-	-	-

In accordance to ASTM C618, the glass satisfies the basic chemical requirements for a pozzolanic material. However, it dose not meet the optional requirement for the alkali content because of high percentage of Na_2O.

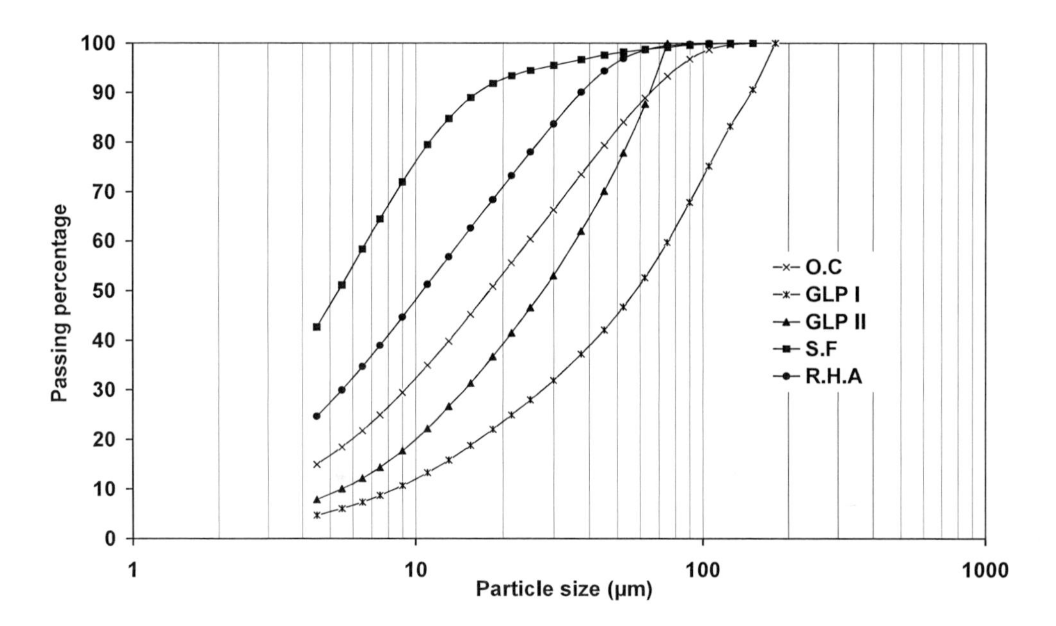

Figure 1. Laser Particle size distribution of ground waste glass type I, type II, silica fume, rice husk ash and ordinary Portland cement.

To satisfy the physical requirements for fineness, the glass has to be ground to pass a 45μm sieve. This was accomplished by crushing and grinding of glass in the laboratory, and by sieving the ground glass to the desired particle size. To study the particle size effect, two different ground glasses were used:

- Type I: ground glass having particles passing a #80 sieve (180μm);
- Type II: ground glass having particles passing a #200 sieve (75μm).

The particle size distribution for two types of ground glass, silica fume, rice husk ash and ordinary Portland cement were analyzed by laser particle size set and have shown in Figure 1.

The particle shapes of all materials were also analyzed using scanning electronic microscope, Figure 2. As shown in Figure 1 and Figure 2 silica fume has the finest particle size. According to ASTM C618, the 180μm and 75 μm glass did not qualify as a pozzolan due to the coarse particle size. Also glass type I and II respectively have 42% and 70% fine particles smaller than 45μm. As mentioned before, review literature shows that that if the waste glass is finely ground under 75μm causes pozzolanic behavior.

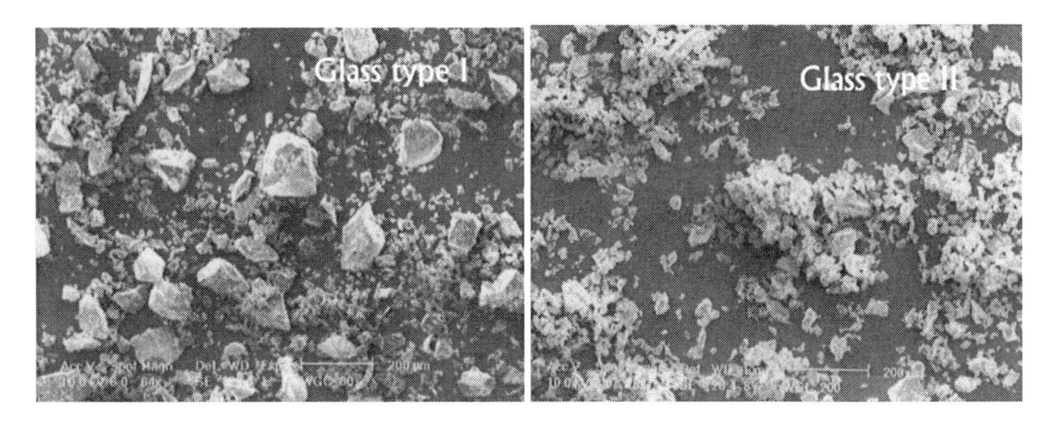

Figure 2. Particle size and shape of ground waste glass type I, type II.

Table 2. Properties of Polypropylene fibers reused in this study

Property	Polypropylene
Unit weight [g/cm3]	0.9 - 0.91
Reaction with water	Hydrophobic
Tensile strength [ksi]	4.5 - 6.0
Elongation at break [%]	100 – 600
Melting point [°C]	175
Thermalconductivity [W/m/K]	0.12

3. Mix Design

Twenty five mixes that contain different fiber volume fraction and different pozzolans were fabricated and tested .the mixture design properties are listed in Table 3.

Table 3. Mix properties

No.	G/C	S/C	W/C	% Content (by weight)					SP/C	%PP	No.	G/C	S/C	W/C	% Content (by weight)					SP/C	%PP
				O.C	GI	GII	SF	RH							O.C	GI	GII	SF	RH		
1	2.80	1.35	0.40	100	-	-	-	-	0.01	0.00	14	2.80	1.35	0.40	90	-	10	-	-	0.01	0.50
2	2.80	1.35	0.40	100	-	-	-	-	0.01	0.10	15	2.80	1.35	0.40	90	-	10	-	-	0.01	1.00
3	2.80	1.35	0.40	100	-	-	-	-	0.01	0.30	16	2.80	1.35	0.40	90	-	-	10	-	0.01	0.00
4	2.80	1.35	0.40	100	-	-	-	-	0.01	0.50	17	2.80	1.35	0.40	90	-	-	10	-	0.01	0.10
5	2.80	1.35	0.40	100	-	-	-	-	0.01	1.00	18	2.80	1.35	0.40	90	-	-	10	-	0.01	0.30
6	2.80	1.35	0.40	100	-	-	-	-	0.01	0.00	19	2.80	1.35	0.40	90	-	-	10	-	0.01	0.50
7	2.80	1.35	0.40	90	10	-	-	-	0.01	0.10	20	2.80	1.35	0.40	90	-	-	10	-	0.01	1.00
8	2.80	1.35	0.40	90	10	-	-	-	0.01	0.30	21	2.80	1.35	0.40	90	-	-	-	10	0.01	0.00
9	2.80	1.35	0.40	90	10	-	-	-	0.01	0.50	22	2.80	1.35	0.40	90	-	-	-	10	0.01	0.10
10	2.80	1.35	0.40	90	10	-	-	-	0.01	1.00	23	2.80	1.35	0.40	90	-	-	-	10	0.01	0.30
11	2.80	1.35	0.40	90	-	10	-	-	0.01	0.00	24	2.80	1.35	1.40	90	-	-	-	10	0.01	0.50
12	2.80	1.35	0.40	90	-	10	-	-	0.01	0.10	25	2.80	1.35	0.40	90	-	-	-	10	0.01	1.00
13	2.80	1.35	0.40	90	-	10	-	-	0.01	0.30											

Study of the expansion due to the possible reaction between the alkali in the cement and the silica in the aggregates was done in accordance with ASTM C1260. The 25 x 24 x 200-mm concrete bars were made of standard graded river sand and gravel, Type 1 Portland cement, and a mineral additive. The water-to-cementitious ratio was 0.4 and twenty batches containing mineral additives, 10% by weight of the Portland cement was replaced by the silica fume, rice husk ash, glass type I and glass type II, respectively. Also for finding the effects of polypropylene fiber to suppress the harmful impacts of alkali-silica reaction lots of mixes were reinforced by different volume fraction of PP fibers (0.1-1.0 %Vol).

All mixes were prepared in a drum mixer with a capacity of 0.15 m^3. Mixing sequence was as follows:

- Place the sand and gravel in the mixer and start the mixer.
- Add one-third of the mixing liquids (water + HRWRA) to the running mixer.
- Add the cementitious materials (cement + pozzolans) slowly with another one-third of the mixing liquids to preserve fluidity and workability of the mixture.
- Add the polypropylene fibers.
- Add the remaining mixing liquids.
- Continue mixing for 5 min, stop the mixer for 3 min, and then continue mixing for an additional 2 min.

After 24 h of curing, the bars were placed in water at 80 °C for another 24 h to gain a reference length. They were then transferred to a solution of 1 N of NaOH at 80 °C. Readings were then taken every day for 20 days. The concrete bars containing different amount of fiber and additives were tested. Also some specimens without any fiber or additives were tested as controls. The comparison with the control is an indication the pozzolanic ability of additives to suppress the ASR and from the solution. It also manifests if the mineral additives used are able to suppress the expansion by consuming more lime in concrete.

4. Results and Discussion

4.1. Analysis of the Expansion Characteristics

The expansion–time history curve made by the measurements complying with ASTM C1260 in terms of the additives like silica fume, rice husk ash and tow kind of ground waste glass whit different particle size which replaced by 10% of cement Portland weight are shown in figure 3.

According to figure specimens containing ground waste glass type I showed higher expansion rates than the other specimens specially the controls ones. The most important reasons for increase in expansion of samples containing glass type I is because of coarse particle size. In this way the alkali content cement and pore water has grater opportunity to react with silica content in aggregate specially particles if glass type I.

By replacing the cement Portland with ground glass type II the expansion rate of the specimens decrease around 15%. According to last researches materials with pozzolanic characteristics can suppress the alkali-silica reaction (ASR) and decrease the expansion respectively. According to ASTM C618 glass type II has the pozzolanic behavior due to its fine particle size.

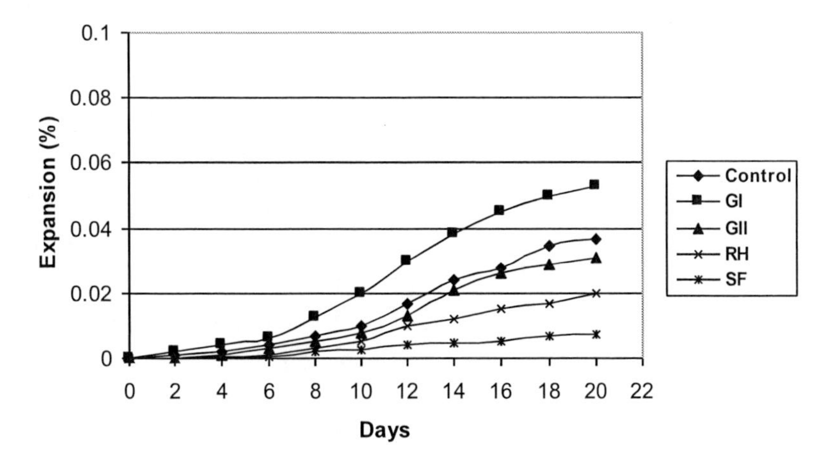

Figure 3. Expansion time history for concrete bars.

By using of rice husk ash in matrix of concrete the amount of expansion decline about 50% of controls. It shows that rice husk ash has more pozzolanic properties the glass type II.

Also application of silica fume instead of 10% Portland cement weight has the most effective impact on expansion of specimens.

Results of mixes incorporating SF showed significantly lower expansion levels at all ages. Specimen incorporating SF expanded 70% less than Control at the end of test.

4.2. ASR Characteristics of Composite Containing Reinforcing Fiber

The internal pressure created by ASR gel causes ASR expansion and cracks in the concrete containing waste glasses.

Such internal expansion can be suppressed by randomly distributing discontinuous single fibers. The ASR expansion was analyzed by varying reinforcing Polypropylene fibers (PP fibers) and different composite cement containing different pozzolans and additions.

Figure 4 display the expansion time history curve for concrete bars having different contents of PP fiber (0.1–1 Vol %), respectively, combine with each type of composite cement.

Results show that the ASR expansion decreases with an increase in the contents of PP fibers. With a steel fiber content of more than 0.5 Vol %, the relative expansion rate defined by ASTM C 1260 was suppressed.

In all cases by increasing the volume fraction of PP fiber in matrix, expansion of specimens decreases. The expansion suppression effect of the reinforcing fiber was more prominent in the glass type I.

At first case with composite cement of glass type I, application of 0.1-0.3 Vol% of PP fiber cause 20% reduction in expansion, but by increasing the amount of fiber to 0.5% and 1% the reduction was around 50% and 70% respectively.

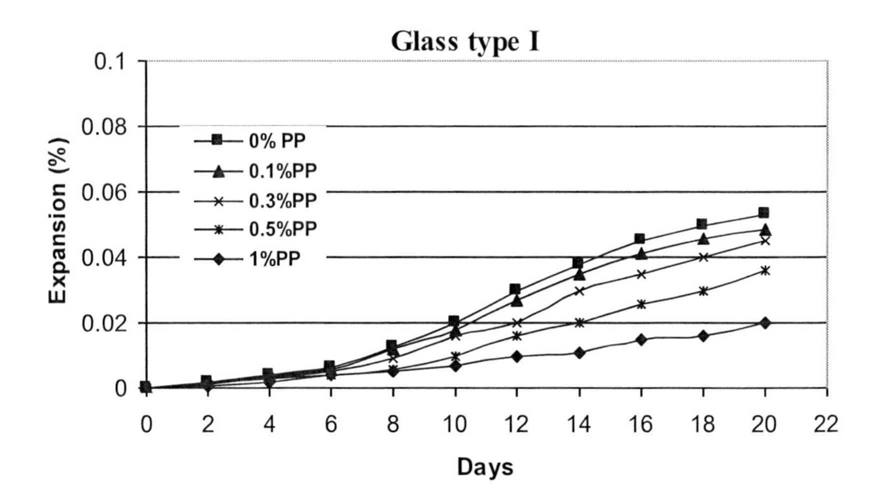

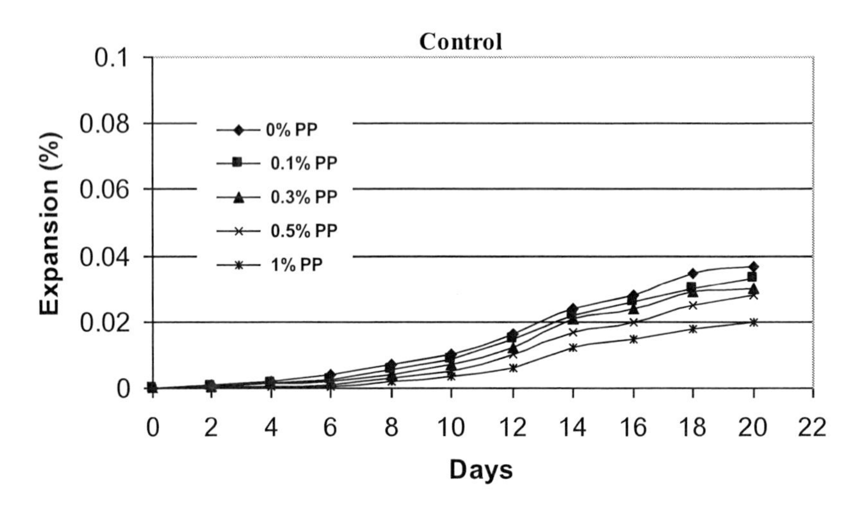

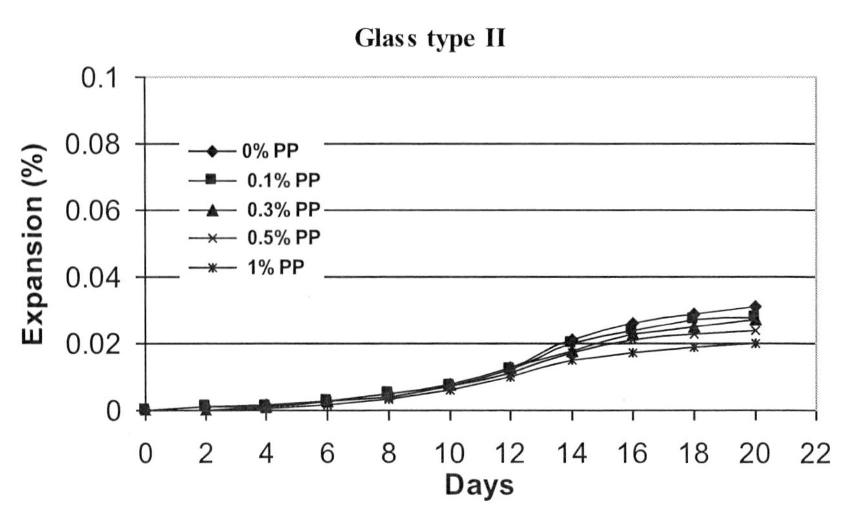

Figure 4 Continued

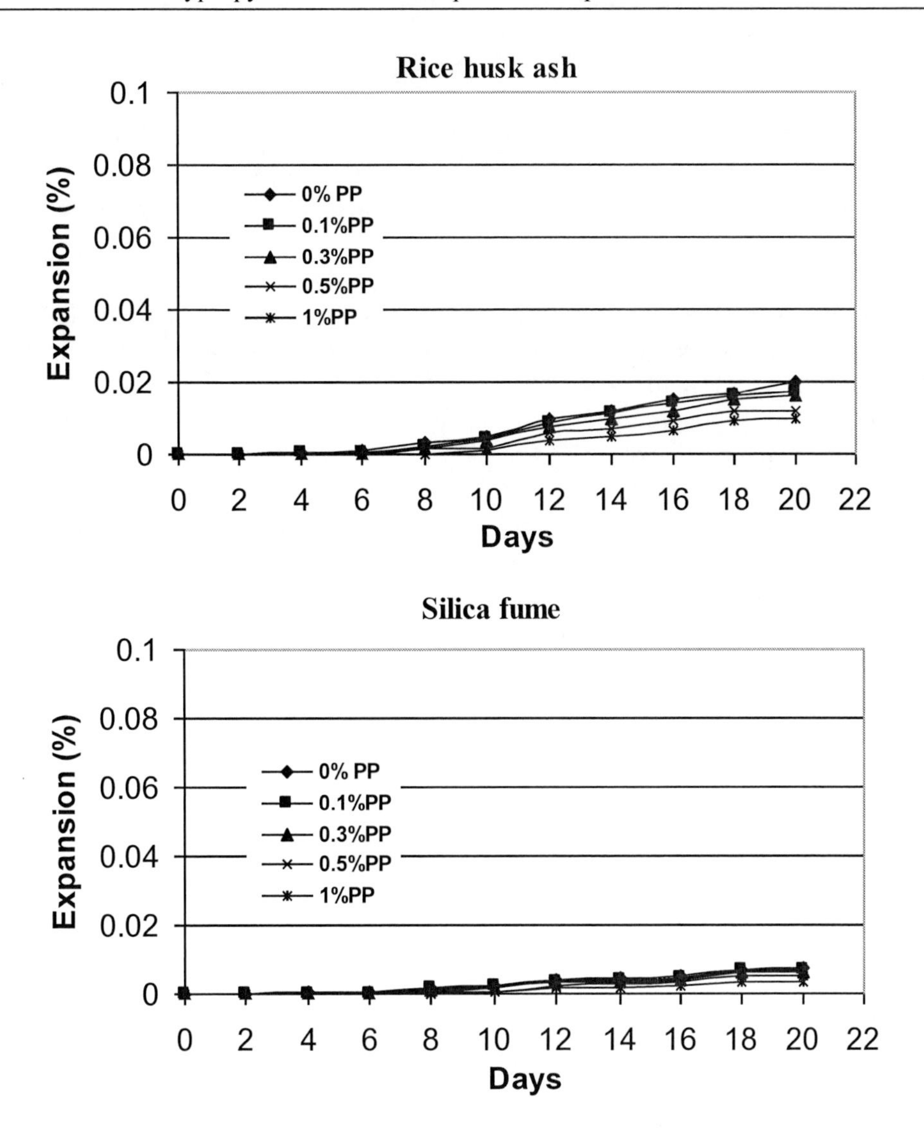

Figure 4. Expansion time history curves for concrete bars containing different amount of PP fibers.

The reduction of expansion in controls was like the specimens with glass type I but it was less. Moreover rice husk ash and glass type II approximately indicates the same results to target specimens. The specimens which contains silica fume (SF) and PP fiber have shown greater reduction in expansion than the others.

In general, results indicate that silica fume, rice husk ash and glass type II have an appropriate potential to apply as a part replacement in cement due to their respective pozzolanic activity index values (according to ASTM C618 and C989, and Table 1) and the high pozzolanic characteristics of these materials have an effective impacts on reduction of expansion caused by ASR.

In this way the lowest expansion belongs to the specimens containing silica fume with PP fibers .the expansion of these specimens was negligible.

Conclusion

This study intended to find effective ways to reutilize the hard to recycle waste glasses as concrete aggregate. Analysis of the problematic ASR expansion of concrete containing recycled waste glasses, silica fume and rice husk ash and reinforcing fiber gave the following results:

(1) Results show that the expansion rate by ASR in accordance with ASTM C 1260 showed an increasing tendency in specimens containing waste ground glass type I due to its coarse particle size rather than the others.
(2) Utilization of high reactive pozzolans like silica fume and rice husk ash have a great impacts on reduction of ASR expansion.
(3) According to results it is obtained that waste ground glass type II shows pozzolanic behavior like silica fume and rice husk ash, subsequently causes the decline in ASR expansion due to its fine particle size according to ASTM C618.
(4) With the content of waste glass type I and that of PP fiber 0.1-0.3 Vol %, the expansion rate of the specimens decreased by about 15-20%. By increasing amount of fiber to 0.5-1 Vol % , the expansion declined around 50-70%.
(5) In general application of fiber in the matrix is a suitable way to suppress the ASR expansion.
(6) It seems that the composition of high reactive pozzolans and PP fiber is the best way to decrease the harmful effects of ASR expansion.

References

[1] J. Refined, Development of non-traditional glass markets, *Resour. Recycl.* (1991) 18–21.
[2] J. Uchiyama, Long-term utilization of the glass reasphalt pavement, *Pavement* (1998) 3– 89.
[3] C.D. Johnson, Waste glass as coarse aggregate for concrete, *J. Test. Eval.* 2 (5) (1974) 344– 350.
[4] O. Masaki, Study on the hydration hardening character of glass powder and basic physical properties of waste glass as a construction material, *Asahi Ceramic Foundation Annual Research Report,* 1995.
[5] S.B. Park, Development of Recycling and Treatment Technologies for Construction Wastes, Ministry of Construction and Transportation, Seoul, 2000.
[6] R.N. Swamy, The Alkali– Silica Reaction in Concrete, Van Nostrand Reinhold, New York, 1992.
[7] S. Naohiro, The strength characteristics of mortar containing glass powder, *The 49th Cement Technology Symposium*. JCA, Tokyo, (1995) 114–119.
[8] I. Kyoichi, K. Atobumi, Effects of glass powder on compressive strength of cement mortar, College of Engineering, Kantou Gakuin University, *Study Report* 40 (1) (1996) 13–17.
[9] Bentur A, Mindess S. *Fiber reinforced cementitious on durability of concrete*. Barking: Elsevier, 1990.

[10] Balaguru PN, Shah SP. *Fiber reinforced cement composites*. New York: McGraw-Hill, Inc, 1992:367p.
[11] S.B. Park, Development of Energy Conserving High Performance Fiber Reinforced Concrete and Instruction for Design and Construction for the Reinforced Concrete, Ministry of Construction and Transportation, Seoul, 1998.

In: Polymer Research and Applications
Editors: Andrew J. Fusco and Henry W. Lewis

ISBN: 978-1-61209-029-0

Chapter 3

BIODEGRADATION OF FILM POLYMER COATING ON THE BASIS OF CHITOSAN

***E. I. Kulish*[1,*]*, V. P. Volodina*[2]*, V. V. Chernova*[1]*, S. V. Kolesov*[1]* and G. E. Zaikov*[3]**

[1]The Bashkir State University, Ufa, Russia
[2]The Institute of Organic Chemistry of the Ufa Scientific Centre, The Russian Academy of Science, Ufa, Russia
[3]Institute of Biochemical Physics, Russian Academy of Sciences, Moscow, Russia

ABSTRACT

The present paper deals with enzymatic biodegradation of film chitosan coatings which can be used for protecting open wounds (burning, surgical ones) and the means of their modification for extending the service life.

Keywords: chitosan, polysaccharide, enzymatic destruction, collagenase, lirase, intrinsic viscosity.

One of the advantages of using chitosan (ChT) for producing various medical materials – films, threads, coatings, gels – is its capability for biodegradation. However, rapid biodegradation under the influence of enzymes may become the factor limiting the usage of such materials by reducing their service life under the conditions of medical application. The present paper deals with enzymatic biodegradation of film chitosan coatings which can be used for protecting open wounds (burning, surgical ones) and the means of their modification for extending the service life. To provide the prolongation of ChT action there are at least two possible ways. The first one is to change its supermolecular structure, to increase the packing density and correspondingly to decrease ChT units accessibility for enzymes. The second way is the transformation of ChT into water- insoluble form.

* E-mail: alenakulish@rambler.ru

Chitosan (TU 6-09-05-296-76) produced by the company "Chimmed" (Russia) was used as the object of investigation. It is produced by deacytelation of crab chitin, the degree of deacytelation being ~75% and M_η=120000. Polyvinyl alcohol (PVA) of the 16/1 brand with M_η=25000 was also used. The enzyme preparation "Collagenase KK" obtained from hepatopacreas of industrial species of crab and produced by SUE "Immunopreparat" (Ufa) was used as enzyme. ChT films as well as films of its mixtures with PVA were obtained by the method of casting of the polymer solution onto the glass surface. Acetic acid (AcOH) was used as the solvent, its concentration varying from 1 to 70% of mass. The polymer mass concentration in the initial solution was 2% of mass. The film thickness in all the experiments was maintained constant and equal to 0.1 mm. The films were dried until reaching constant weight. For modelling the process of biodegradation of a ChT film specimen on wound surface the film was placed on the base moistened with collagenase solution and was held at 35°C for 1.5 hour. The biodegradation degree was determined according to the data of viscosimetry in the buffer solution consisting of 0.3 M AcOH and 0.2 M sodium acetate. The transformation into the basic form was carried out by holding the film in 0.1 N NaOH solution for 1 hour with subsequent washing in distilled water. The thermomodification of films was carried out by heating at 120°C. The swelling of film specimens was determined by the gravimetric method.

The principal possibility of ChT biodegradation process under the influence of the enzymatic complex "Collagenase KK" has been established earlier [3]. In addition to ascertaining the destruction fact itself it has been found out that the conformational and supermolecular state of ChT in the film and correspondingly the rate and degree of ChT biodegradation are greatly determined by its aminogroups protoning degree which in its turn depends on the concentration of the acid used. The variations in ChT state in AcOH solutions are demonstrated by the values of intrinsic viscosity of ChT which hasn't been in contact with enzyme yet (Figure 1, curve 1).

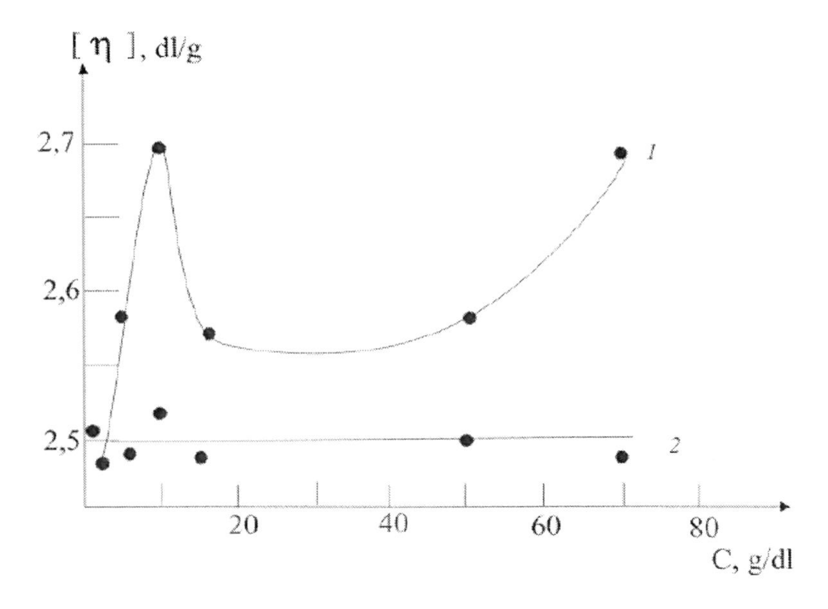

Figure 1. The dependence of the limit value of viscosity of ChT films in the salt (1) and basic (2) forms which were not subjected to biodegradation, on the concentration of acetic acid in initial solution.

For all ChT films the type of dependence of viscosity variation on the time of holding in contact with enzymatic complex is the same (Figure 2).

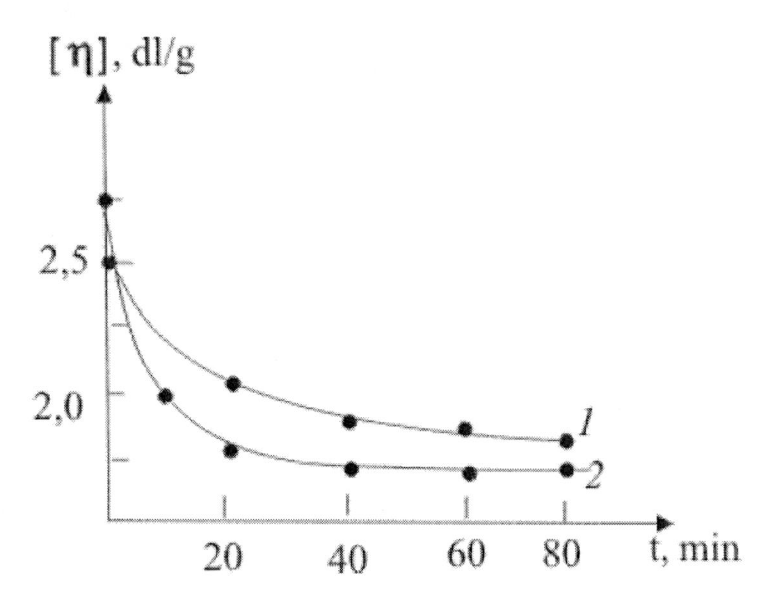

Figure 2. The dependence of the limit value of viscosity of ChT solutions obtained from film specimens in the salt form made of 1% (1) and 10% (2) acetic acid on the time of contact with enzyme.

The most significant drop of intrinsic viscosity takes place in the initial (15-20 min) period. Further enzymatic exposure of specimens doesn't greatly affect the degree of intrinsic viscosity drop. One can also note that both at short and at long periods of holding the films in contact with enzyme solution one and the same trend persists – the least variation of polymer molecular mass is observed in the film obtained from 1% acid while the greatest one – in the films from 10% and especially 70% AcOH (Table 1).

Table 1. The characteristics of ChT film specimens obtained from initial solution with different acetic acid content. (The time of holding in collagenase solution is 1 hour. Molecular mass of initial chitosan is 126.*10^5 (c [η]=3.0,dl/g).

Δ[η] in the process of biodegradation, dl/g	Concentration of AcOH in the initial solution, dl/g.						
	1	5	10	15	20	50	70
For salt films	1.03	1.28	1.33	1.15	1.08	1.12	1.59
For basic film	0.51	0.46	0.63	0.51	0.60	0.65	0.72

ChT films obtained in the salt form (ChT acetate) have good solubility in water and high moisture-absorbing ability (Figure 3). High moisture-absorbing ability of ChT is its undeniable advantage, however, good solubility of such films promotes the accelerated mechanical destruction of a polymer film on wound surface.

Chitosan films holding for 1 hour in 0.1N NaOH solution is accompanied by ChT transformation from the salt form into the basic one [4]. This transformation results in a considerable increase in the density of ChT macromolecules packing with a corresponding decrease in the degree of accessibility of chitosan units for interaction with enzyme.

The variation of chemical and supermolecular ChT structure in the films transformed into the basic form results in a number of consequences. First, the transition of ChT films into the basic form is accompanied by a considerable decrease in, the degree of film specimens swelling (Figure 3, curve 2). Second, there occurs levelling in the values of ChT intrinsic viscosity (from the comparison of curves 1 and 2, Figure 1).

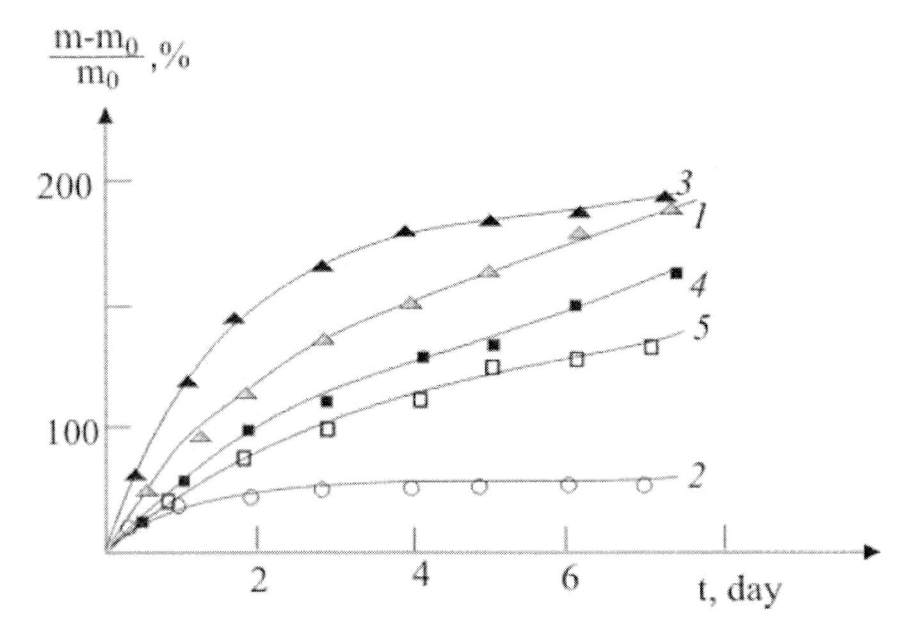

Figure 3. The kinetics of swelling in water vapours of ChT film specimens obtained from 1% acetic acid in the salt form (1), basic form (2), subjected to thermal treating (3) as well as the mixtures of ChT-PVA of the composition 80:20 (4) and 50:50 (5).

This situation differs significantly from that which was observed in the case of films in the salt form where spread in viscosity values is great. Third, the general form of the curve of dependence of viscosity variation on the time of film contact with solution of the enzyme complex "Collagenase KK" (Figure 4) is the same as in the case of films obtained in the salt form, however, the degree of film specimens biodegradation is considerably lower (see Table 1).

As in the case of biodegradation of salt films the greatest degree of decreasing ChT molecular mass is observed in the case of basic films obtained from 70% AcOH but the general degree of biodegradation of ChT obtained in the basic form is considerably less than that of ChT in the salt form.

Significantly smaller extent of enzymatic degradation of basic films as compared with salt ones is their undeniable advantage because it provides the possibility of longer stay of the film on wound surface without destruction. However, mechanical properties of such films are not high: they are fragile, not strong enough and non-elastic. Besides, such films have low moisture-absorbing ability which suggests their low ability to absorb the liquid given off by wounds and doesn't make these films very suitable for practical use.

Considerable improvement of mechanical properties of the film polymer coating may be attained by adding another water-soluble polymer-PVA- to Chitosan at the stage of film formation. In this case the films formed are strong and elastic [5, 6]. It should be noted that under the chosen conditions PVA doesn't biodegrade. This is demonstrated by the fact that its relative viscosity and intrinsic one are constant for a long period of its contact with enzyme.

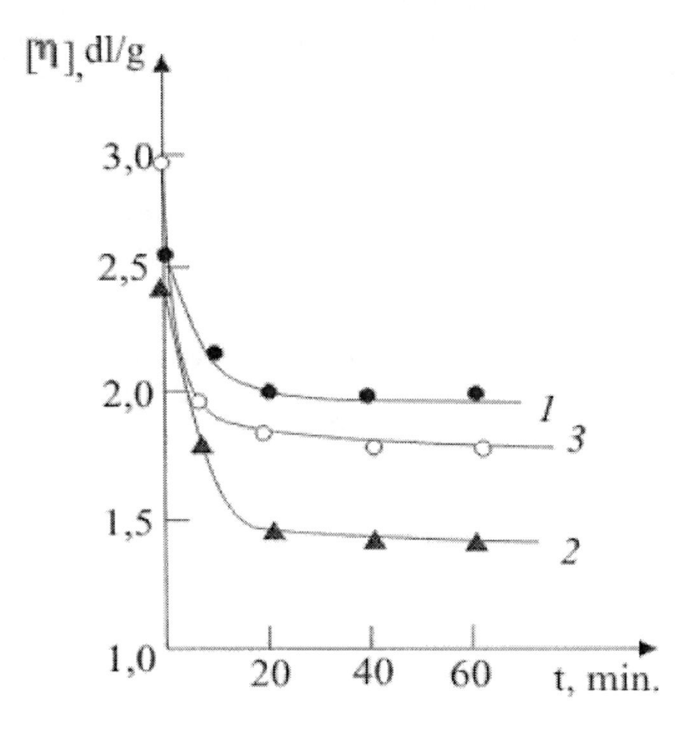

Figure 4. The dependence of the limit viscosity value of solutions of ChTobtained from film specimens formed from 1% acetic acid in the basic (1) and salt forms (2) and heated for 1 hour at 120°C, on the period of contact with enzyme.

However, in the case when the ChT-PVA specimens obtained in the form of films from 1% solution in AcOH begin to contact with enzyme, the ChT enzymatic destruction at all the investigated polymer relationships in the mixture takes place in a greater extent than for individual ChT (Table 2).

Table 2. The characteristics of ChT-PVA film specimens obtained from solutions in 1% acetic acid (The time of holding in collagenase solution is 1 hour. The molecular mass of initial chitosan is $1.26*10^5$ ([η] =3.0 dl/g))

Variation in the process of destruction	PVA content in the mixture, % mas.				
	0	20	40	60	80
Δ[η] , dl/g	1.03	1.32	1.35	1.35	1.40
$\Delta M_\eta*10^{-3}$	36	48	50	50	52

As mentioned in [7] irrespective of acid concentration in the initial solution, in the formed ChT-PVA film there takes place a decrease in the number of supermolecular formations and an increase in their size as compared with additive values. This fact is evidently due to the existing interpolymer interaction, i.e. the formation of ChT-PVA heteroaggregates which are known to be characterized by lower packing density and by great sizes. Thus, the introduction of the second polymer is likely to change the supermolecular structure of the initial film, which is embodied in the kinetics of biodegradation of the film polymer composition. ChT accessibility for its interaction with collagenase in such loosely-packed aggregates is greater than in the case of ChT-ChT homoaggregates with high packing density. It should be noted, however, that ChT-PVA film polymer coatings don't have high sorption ability despite their good solubility in water because individual PVA doesn't possess high moisture-absorbing ability (Figure 3, curves 4, 5).

To obtain films insoluble in water which preserve a high degree of swelling is possible by thermal modification of films. As known [8] the heating of chitosan films for 3 hour at the temperature about 120°C is accompanied by significant rearrangement of the chemical structure, which doesn't, however, affect their swelling ability (Figure 3, curve 3). Since in the case of films insoluble in water it isn't possible to analyze the capacity for destruction, the specimens put through thermal treatment for 1 hour during which films still retain the capacity for dissolution in water, were taken for investigation. Nevertheless, in this case one can observe a pronounced tendency for increasing ChT intrinsic viscosity depending on the time of specimens heating (Figure 5, curve 1). The evident reason of increasing intrinsic viscosity is the process of ChT joining accompanying the modification. At increasing the time of thermomodification the film stability to the process of biodegradation also increases (Figure 5). In fact, biodegradation of such thermomodified specimens proceeds up to reaching the viscosity values close to those which they have in basic films as well (Figure 4, curve 3). However, such thermomodified specimens, unlike basic films, are characterized by good mechanical properties.

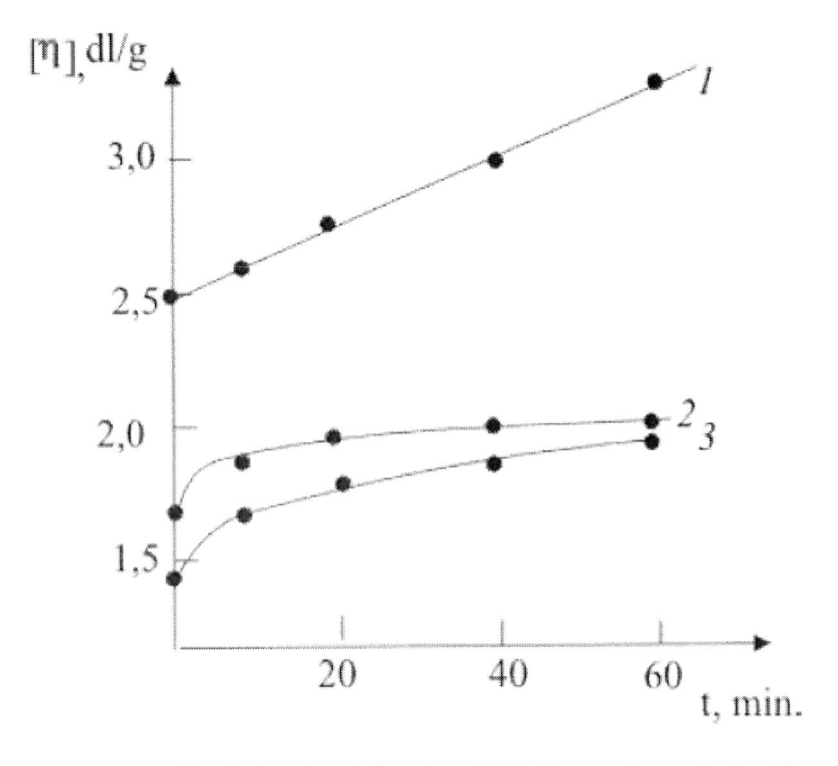

Figure 5. The dependence of the limit value of viscosity of ChT film specimens obtained from solution in 5% acetic acid on the time of heating at 120°C. 1- the films not subjected to contact with enzyme, 2- the films contacting with enzyme for 10 min, 3- the films contacting with enzyme for 30 min.

Thus, the chemical modification of chitosan films carried out by treating the formed film with alkaline solution or by thermal treatment (joining) of the film makes it possible to obtain polymeric coating with a good complex of physical properties and increased stability to enzymatic degradation.

Thus, the following conclusions can be made from the given work:

1. Thermomodified chitosan films are characterized by the absence of solubility in water, high moisture-absorbing ability, high strength and elasticity. At increasing the time of thermal treatment the stability of chitosan films to the process of enzymatic biodegradation also increases.
2. Modified films in which ChT is in the basic form are characterized by low solubility in water as compared with non-modified films, and by low moisture-absorbing ability. The given films possess high stability to the process of enzymatic biodegradation and slightly depend on the prehistory of the film formation.
3. Mixture polymeric coating "chitosan-polyvinyl alcohol" have high strength and elasticity, increased biodegradating ability and good solubility in water. Moisture-absorbing capacity of such films is somewhat lower than that of individual chitosan.

References

[1] Chitin and Chitosan: Obtaining, properties and use. /Under the red. of Scraybyn K.G., Vikhoreva G.A., Varlamov V.P. M.: *Nauka*, 2002, (in Russian).

[2] Muzzarelli R.A.Chitin.Oxford : Pergamon Press, 1977, 309 p.

[3] Kulish E.I., Volodina V.P., Fatcullina R.R., Kolesov S.V. Proceedings of the 8 th International Conference "*Modern prospects in the Investigation of Chitin and Chitosan*". M.,VNIRO. 2006. S.290.

[4] Ageev E.P., Vikhoreva G.A., Matushkina N.N., Pchelko O.M. // *Vysokomolec. soed.* A. 2000. V.42. №2. S.333.

[5] Mukhina V.R., Pastukhova N.V., Semchikov Yu.D., Smirnova Z.A., Kiryanov K.V., Zhernenkov M.N. //*Vysokomolec. soed.* 2000. V. 43 A. №10. S.1797.

[6] Nikolaev A.F., Prokopov A.A., Shulgina E.S., Golenyscheva S.A., Klubikova Z.E., Musikhin V.A. // *Plast.massy*.1987. №11, S.40.

[7] Kulish E.I., Kolesov S.V.. // *J. P. Ch*.2005.№9. S.1511.

[8] Zotkin M.A., Vikhoreva G.A., Kechekyan A.S. // *Vysokomolec. soed.* B. 2004. V.46. №2. S.359.

In: Polymer Research and Applications
Editors: Andrew J. Fusco and Henry W. Lewis
ISBN: 978-1-61209-029-0

Chapter 4

GRANULATING OF REDUCED TO FRAGMENTS SECONDARY POLYMERS BY MEANS OF EXTRUSION HEAD

T. G. Beloborodova, A. K. Panov, T. A. Anasova and G. E. Zaikov
Academy of Science of Republic of Bashkortostan,
Sterlitamak Branch, Sterlitamak, Russia
N.M. Emanuel Institute of Biochemical Physics,
RAN, Moscow, Russia

ABSTRACT

The description of the construction, the principle of action and calculation method of geometric parameters of extrusion head for granulating to reduced to fragments polymeric waste materials are stated in this work. Experimental data of the process of granulating are described here too.

Keywords: Extrusion head, extrudate, granules, draw plate, forming channel, screw.

INTRODUCTION

Steady rise of volume of polymeric production and expansion of usage of polymeric materials is inevitably accompanied by accumulation of industrial and domestic plastic waste materials, which lead to the increase of economic and ecological problems.

In this connection, processing of waste materials of polymeric items, that is sleeve pellicle tubes, cords, plaits, bottles, different profile items and other plastic items acquires more and more importance according to the point of view of saving material resources and solving of ecological problems. One of the trends of waste material utilization is their processing into polymeric raw materials and its secondary usage for receiving different items.

Waste materials of polymeric items is necessary to reduce to fragments beforehand, and then to granulate [1]. Granulated secondary raw materials are in the shape of granules with standard grains, constant volume mass and some qualities of good dry substances [2]. For reducing to fragments of polymeric waste materials, we have worked out and got a patent on a plant [5] for reducing to fragments of "soft" polymeric waste materials. It is an independent unit capable to process a wide range of polymeric waste materials, has enough productivity, and at the same time is relatively simple in the sense of construction decision.

Extrusion Head Construction Used for Granulating Reduced to Fragments Polymeric Waste Materials

Introduced extrusion head is a forming instrument of shneck press-granulator for granulating reduced to fragments polymeric waste materials. During elaborating of a forming head used for granulating reduced to fragments polymeric waste materials it was aimed to achieve uniform outlet of extrudate out of draw plate and improvement of quality of granules, which are received from reduced to fragments secondary polymers. This task is solved by several constructive peculiarities of forming head.

Extrusion head used for granulating reduced to fragments secondary polymers is comprised of a body with a feeding channel for feeding reduced to fragments materials, a pressing flange and a draw plate with a forming channel of variable length and a prominent profile of cross section of draw plate directing to the side of a feeding channel and is being made in the central part as an ellipsoid. It passes on to the periphery of gently sloping curve as one cavity giperboloid, and the ratio of length of forming channels of a draw plate to the diameter of its prominent surface within 0.15-0.35 and has its maximum in the centre of a draw plate.

The result provided by extrusion head used for granulating reduced to fragments secondary polymers is expressed in quality rise of granules thanks to the choice of optimum geometry of cross section of a draw plate and a ratio of length of forming channels of draw plate to the diameter of its prominent surface within 0.15-0.35.

If this ratio is less than 0.15, then granules when forming do not possess enough density and do not have smooth enough surface. Extrudates become rough and when cutting, they are crumbled even being insignificantly influenced mechanically.

When ratio is from 0.15 to 0.35, granules have enough solidity and quality.

When thickness of draw plate enlarges and consequently length of forming channels rises more than 35% from the diameter of its prominent surface hydraulic resistance of extrusion head and this lows the plant productivity.

Granulated material must have size exactness of granules. Speed of flowing out of extrudate is not the same in cross section of channel: in the centre of draw plate, it is higher than in the periphery that is why for size exactness the uniformity of their outlet must be provided.

In this case leveling of linear speeds of extrudate outlet is provided by deceleration of hydraulic resistance on perypherical areas of draw plate because of the length of forming channels is lessened from the centre of a draw plate to its periphery. Draw plates are made in the form ellipsoid and in the periphery; they are in the shape of gently sloping curve as one cavity giperboloid. The form of a draw plate surface in mentioned areas in each case is sorted out constructively depending on the type of the material being granulated.

Cross section of extrusion head used for granulating reduced to fragments secondary polymers is shown on figure 1. Eextrusion head used for granulating reduced to fragments secondary polymers is comprised of a body 1,a draw plate, a pressing flange 3, which is fixed to the body by bolting junction 9. Forming channel of head is formed by a cylindrical channel 4 for feeding of reduced to fragments polymer, passing into a cone channel 5 and forming channels 6, made in draw plate 2.

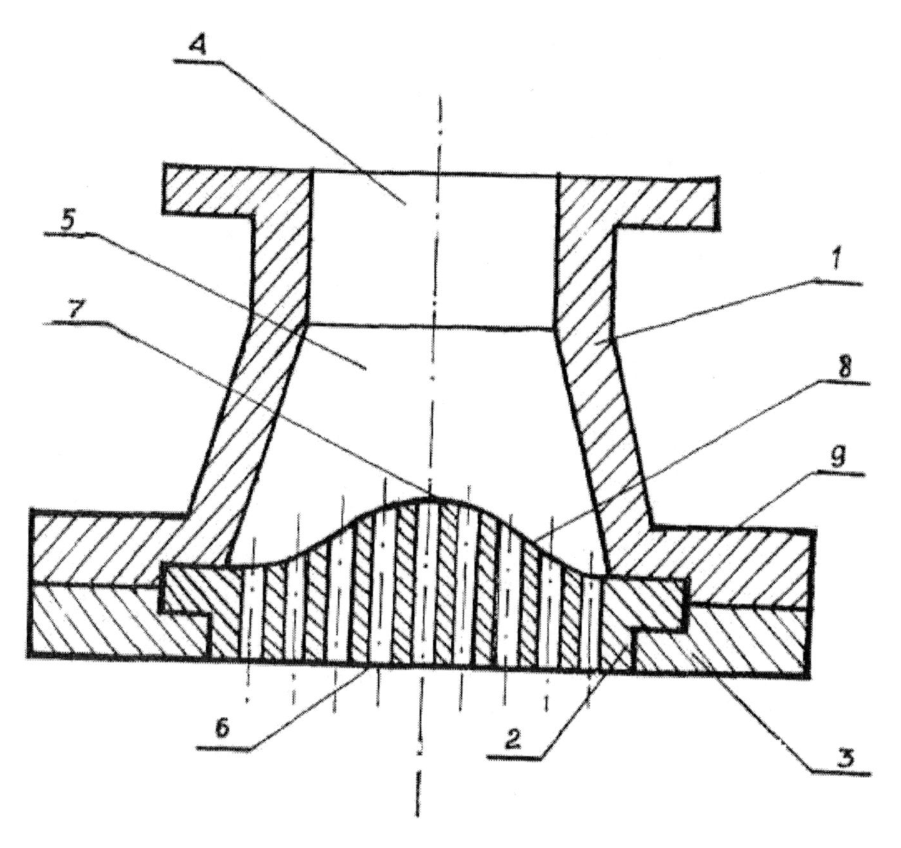

Figure 1. Cross-section of extrusion head.

Prominent surface of draw plate 2 is profiled in its central part as an ellipsoid 7 passing to its periphery as a gently sloping curve on the form of one cavity hyperboloid satisfying the following equation:

$$Y=kx^2,$$

where

y- an ordinate of hyperboloid points, mm

k – coefficient of pressing hyperboloid branches, mm

x – an abscissa of hyperboloid points, mm

Forming channels 6 have variable length, which is lessened from the centre of a draw plate to its edges.

In figure 2 a cross section of draw plate with its prominent profile 7 (y , x) and 8 (y , x) and a flat base of draw plate (y , x) are shown; coordinated axes are located according to figure 2. After the process of profiling of prominent profile of draw plate forming channels with diameter 6mm and pace 10mm were drilled.

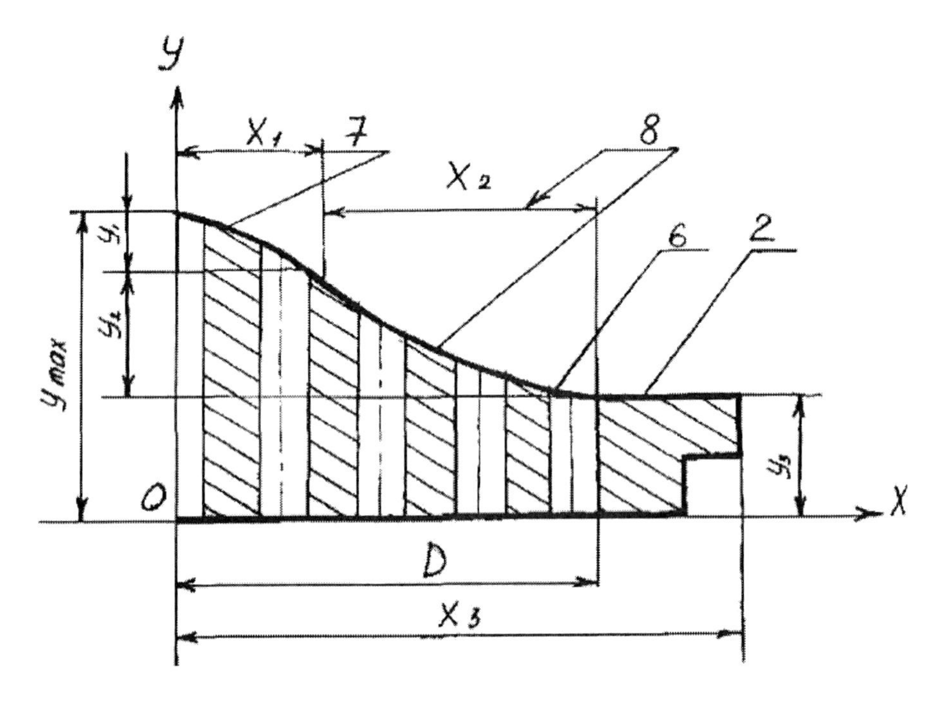

Figure 2. Cross section of a draw plate.

Having known the length Li of forming channels, located on a draw plate, the ratio of forming channel length the diameter of its prominent part Li/D was determined and it was within 0.15 – 0.35.

Experimental Part

The process of granulating of reduced to fragments in extrusion head polymeric materials was researched experimentally. Extrusion head used for granulating reduced to fragments secondary polymers works in the following way.

Reduced to fragments material is given through cylindrical channel 4 into cone channel 5 where it is smoothly distributed alongside its cross section and goes through forming channels 6, forming, at the same time, some extrudate material, which in its turn is cut into granules. Pressure in head's channel is lessened when moving off its centre to periphery. Leveling of pressure in the head channel is provided by prominent profile of cross section of draw plate is executed in the form of ellipsoid in its central part (figure), passing to the periphery gently sloping curve like one cavity hyperboloid 8 according to figure 1.

Experimental research of extrusion head for granulating of reduced to fragments secondary polymers were done on experimental plant with extrudate material ATL -45, which has a shneck with length 45mm and length 1125mm.

Extrusion head had the following size: general diameter 210mm, length 170 mm, diameter of feeding channel – 60mm, diameter of prominent surface -150mm, thickness of central part of draw plate – 60mm, diameter of forming channels of draw plate – 6mm, screw pitch between them-10mm. The process of granulating was executed with reduced to fragments secondary polymers, which was received from sleeve pellicle PVD 15802-020 with T=403-423K and reduced to fragments waste materials cable plasticate O-40 with T=433-433K with extrusion pressure in the range P=0,1-12,0MPa. Solid granules of exact cylindrical form with even smooth surface and identical length were received.

Working Out of Method of Calculation of Granulating Plant

The aim of calculation methods of equipment, used in granulating processes is definition of their pressure- outlay characteristics, which give an idea of not only technical but also technological possibilities of the equipment [4].

Many mathematic investigations are devoted to the processes, which take place in shneck equipment. Their main problem is in the usage of reological ratio of condition of processed materials. Given ratio in quantity, relation must guarantee admissible engineering exactness of prognoses of shneck work equipment.

Such rations still unknown for dry substances. That is why modern principles of calculation for extrusion machines, aiming at polymer processing, were used for mathematic modeling of work of screw granulating plant.

Having analyzed calculation methods of extrusion equipment for production of polymer pellicles, blown and profile – linear items, tubes sheets, and for granulating of polymeric materials, it was stated that modeling of work of extrusion equipment comes to mathematic description of joint interaction of dozing zone of screw machine and forming instrument (extrusion heads of difficult type).

For conducting process of extrusion granulating of reduced to fragments materials, many let out in series screw machines are known. The “weak point” of which is forming head. That is why for calculation of granulating plant we offer method, based on the necessity of defining optimal geometric parameters of forming head (thickness of draw plate, diameter of forming hollows), which comes to decision of ratio system, being pressure – outlay characteristics of screw machine and forming head.

Pressure – outlet characteristic of forming head looks like:

$$Q = -\frac{\pi \cdot n \cdot a \cdot R^{k+3}}{2^k (k+3)} \cdot \left(\frac{\Delta p}{L_i} \right)^k \qquad (1)$$

where Q – volume outlay of granulate, ; a, k – reological constants; R – radius of forming channel of draw plate, m; L – thickness of draw plate, m; Δp – process pressure, Pa; n – number of forming channels.

Radius of forming channels in draw plate was chosen according to demands of concrete technological process, during which received granules are used. Optimal pressure was defined according to the results of investigations of plastic solidness of granulated material; volume outlay was defined by matching of pressure – outlay characteristics of screw machine and forming head. Putting found optimal parameters Q, r and Δp into formula (1) optimal thickness of draw plate L; was got for given material.

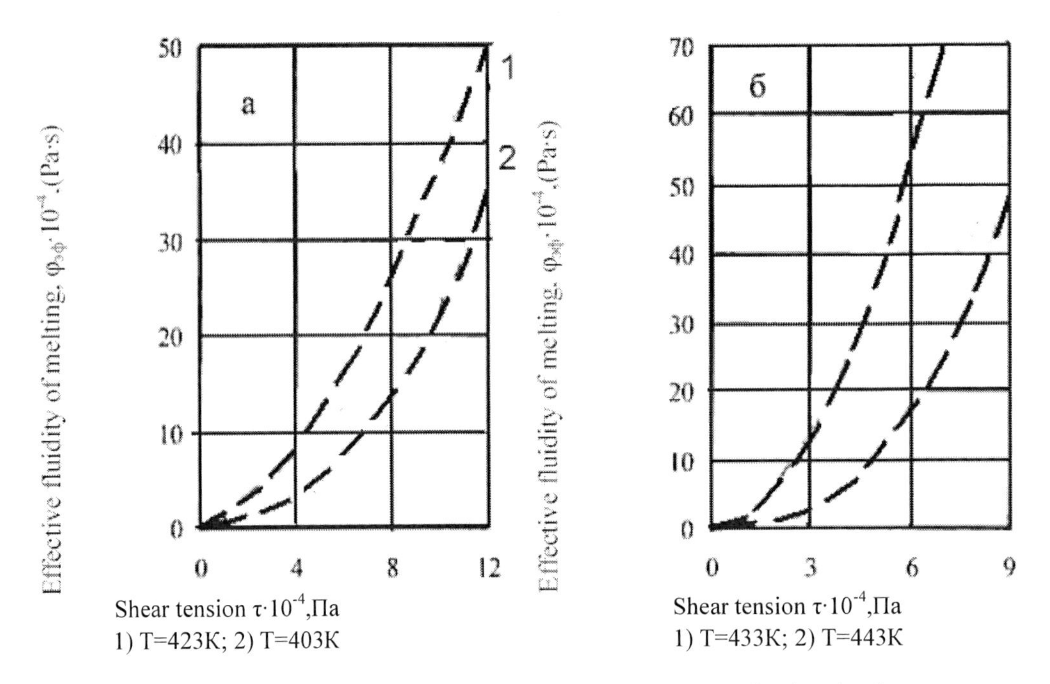

Figure 3. Dependences of fluidity of melted material on tension of shear of reduced to fragments waste materials PVD sort 15802-020 (a) and cable plasticate sort O-40 (b).

Thus, offered method of granulating reduced extrusion equipment for granulating reduced to fragments materials allows to calculate theoretically pressure – outlay characteristic of extrusion equipment, depending on qualities of processed materials, to define optimal granulating rate, i.e. to fulfill calculation of forming instrument.

Elaborated extrusion head for granulating reduced to fragments secondary polymers allows, thanks to choice of optimal geometry of draw plate cross section, to raise quality of received granules.

Dependencies of fluidity of melted materials of polymers on tension of shear of reduced to fragments waste materials PVD, sort 15802 – 020(a) and cable plasticate sort 040(b) are shown on figure 3. Elastic qualifies of melting of waste materials differ insignificantly from properties of primary polymers.

CONCLUSION

1) Extrusion head with hyperboloid draw plate is elaborated. It provides smooth outlet of granulated material.
2) Experimental researches of high – elastic properties of primary polymers and their reduced to fragments waste materials are conducted. Their "degree" character is worked out too.
3) Universal method of calculation of granulating plant for reducing to fragments of polymeric items is elaborated. This method allows defining optimal geometrical parameters of granulating equipment, depending on the properties of granulated material.

REFERENCES

[1] Beloborodova T.G., Panov A.K. Effective constructions of reducers for "soft" polymers processing. Problems of machine – conducting and critical technology in machine – building complex of the Republic of Bashkortostan: Collection of scientific works. – Ufa: "Gilem", 2005.-p.300.

[2] B.B.Bobovich "Utilization of polymeric waste materials": Textbook.-M., 1998 - p. 62.

[3] German H. Screw machines in technology. M,: *Chimiya*, 1972.- p.278.

[4] Lukasik V.A. and others. Basis of industrial processing of polymers. *Basic materials, manufacturing of polymeric compositions: Textbook – Volgograd*, 1997. – p.112.

[5] Patent 2116196, Russia, MKI V2917/00, V 02S18/44. Plant for reducing to fragments of elastic waste materials/ A.K.Panov, Beloborodova T.G. and others - № 96107939/25, Declared 19.04.96; Published 27.04.98, Issue № 21.

[6] Patent 2205104, Russia, MKI V29 V9/06, V29 S47/30. Extrusion head for granulating paste- like materials/ A.K.Panov, I.Kh. Bikbulatov, A.A. Laktionov; Declared 22/03/02; Published 27.05.03, Issue № 15.

In: Polymer Research and Applications
Editors: Andrew J. Fusco and Henry W. Lewis

ISBN: 978-1-61209-029-0

Chapter 5

INCIDENCE OF THE AGGRESSIVE ATMOSPHERIC AGENTS FROM THE CUBAN CLIMATE ON POLYMERIC ARTICLES AND MATERIALS

Orlando Reinosa Espinosa*[*1], Francisco Corvo Perez[2] and G. G. Makarov[3]

[1]Central National Research Center of Cuba, Chemical Division, Habana, Cuba
[2]Institute of Materials and Chemical Compounds for Electronics, Habana University, Cuba
[3]N.M. Emanuel Institute of Biochemical Physics, Russian Academy of Sciences, Moscow, Russia

ABSTRACT

The present research shows the high aggressiveness on the polymeric material and articles and it specially shows the aggressive factors of the humid and tropical climate of Cuba on the rubbers the different and more significative atmospheric factors, that is to say (temperature, solar radiation, ozone concentration) were supplied by the Instituto de Meteorología de la República de Cuba (The Institute of Meteorology of the Republic of Cuba). The samples of the tested rubbers, especially elaborated with no antioxidant protections were made using the coded rubber, used by the Rubber enterprise Union with the number 471.

The evaluation of the spouting and development of the ageing of the rubber samples mounted on the resting frames and under tension was carried out according to prior studies based on internationals standards.

The early apparition of cracks (days of exposition perpendicular to the direction they were mounted on the testing frames is a demonstration of the high aggressiveness of the humid and tropical climate of Cuba.

The results obtained and presently shown in this work ,show, on one side ,to give a good antioxidant protection to the rubber and their articles to be used under the atmospheric condition under the humid, tropical climate of Cuba and also that the

* Orlando.Reinosa@cnic.edu.cu

atmospheric testing stations used could be employed to select antioxidant protection on systems of high protective effectiveness

Keywords: climate, materials, polymers, atmospheric agents, aggressive media.

INTRODUCTION

The Cuban climate favors the initiation and development of certain chemical and physical processes that reduce the useful properties of polymeric articles and materials, mainly those of massive use like rubbers, plastic and composites. These processes occur in very short periods of time in comparison with the durability of similar articles in some other countries with very different kinds of weather.

Although it is known the significant impact that these processes cause on the Cuban economy, but there are not specific dates to reflect them. No evaluation has been made up to the present moment.

Due to the easy appearance of this phenomenon in our country and its undesirable aspects not only from the technical point of view, but also from the economic one, a research is demanded with the aim of knowing all the mechanism that take part in the process and also the request for the appropriate procedures to inhibit it stabilization.

Though a particular research at a certain level is carried out Nowadays a deeper investigation is demanded. The purpose is to obtain a higher durability on our articles that will result on a higher production, durability and better quality.

AGGRESSIVE ATMOSPHERIC AGENTS PRESENT ON THE CUBAN CLIMATE

Table 1. Comparison of bond energy of most common chemicals groups present on polymers and radiation energies

Bond C-C	Bond energy (KJ / mol) 955.0	Radiation Energy, (KJ / mol) 1197.0	Wavelength (nm) 100
C≡N	877.8	-	-
C=O	730.8	798.0	150
C=C	520.8	598.5	200
C-F	499.8	-	-
OH	462.0	478.0	250
C-H	411.8	-	-
Si-O	373.8	-	-
N-H	352.8	399.0	300
C-Cl	327.6	342.0	350
Si-C	294.0	299.0	400

In our country there is a tropical climate with a high humidity and marine influence. Another significant aspect is the condition of island and our location in a region where the influence of cold fronts (north) and tropical storms and hurricanes (south) is received.

Among the determinant factors on the formation of the Cuban climate we may identify the amount of sun radiation that is received, the characteristic of the atmospheric circulation over Cuba and the influence of the different physical and geographical traits our territory.

Table 2. Some atmospherics parameters from two different assay stations

Factors	CNIC station	FLORES station
Humidity (%)	79	79
Solar Radiation mean (MJ / m^2)	17.9	17.9
Solar activity	236.3	--
Temperature mean (°C)	28.0	28.0
Salinity (mg / m2 / day)	21.1	430.5
$S0_2$	21.1	430.5
Dust (mg / m^2 / day)	64.9	--

Effect of the Sun Radiation on Polymeric Materials

Sun radiation has a high impact on polymeric materials and its articles. The most aggressive components are the radiations from UVA region that cause a strong damage not only on materials but also on human beings.

In Cuba the average annual temperature is 25 °C, highest temperatures are around 37 °C and the lowest between 1 °C-10 °C. The hottest region is the eastern one. In the following frame there are some figures reported by two stations in Havana city.

Action of Ozone on Polymeric Materials and Rubbers Articles

Ozone is one of atmospheric factors highly significant on degradation of polymers. It has been demonstrated that low concentrations of ozone in the environment, ground 10-50 microgrames/m^3 in air, are enough to provoke the initiation and development of degradation on polymeric materials mainly on rubbers and as a result of it these rubbers are rejected since they are not good to be use.

In the following frame the differences between the two seasons regarding ozone's concentration have been stated. There is a dry season, which is (winter time), with low temperatures and not very much of rain, but a higher concentration of ozone and on the other hand we have the rainy season (summer period) with high temperatures and a lot of rain, but lower concentration of ozone and effects of sun radiation on polymeric materials.

EFFECTS OF SUN RADIATION ON POLYMERIC MATERIALS

Sun radiation is very significant on the behavior of polymeric materials and articles. The most aggressive components are undoubtedly, the radiations from UV region. These radiations are enough to provoke serious damages not only on materials but also on human beings. In the following diagram, the different behaviors of three rubbers formulations are shown. Every formulation has a different ozone protection and they were exposed at two different regimes: shade and sun. The effect of temperature and sun radiation can be appreciated depending on the exposure conditions. Diagrams reflecting the development of the aging process in rubbers wether they are exposed to shade or to the sun.

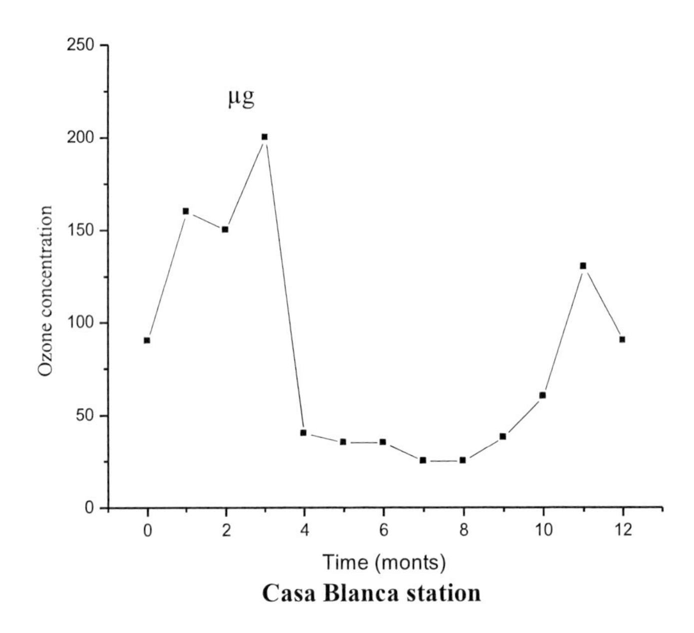

Figure 1. Annual Ozone concentration on Havana City, Cuba.

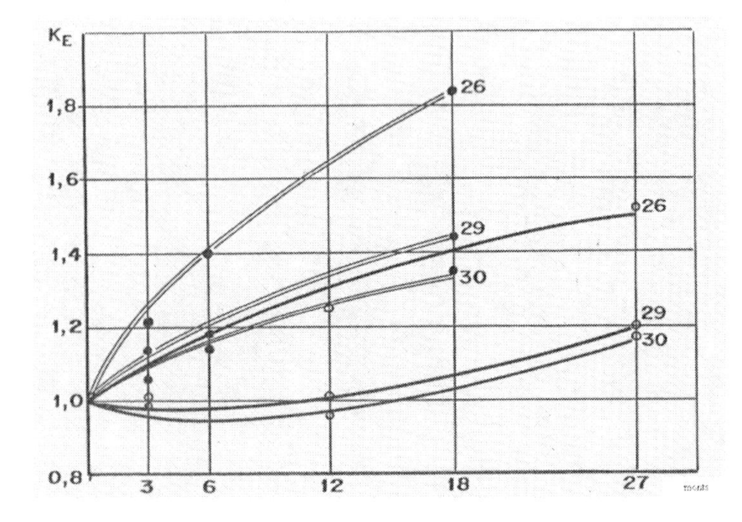

Figure 2. Dependence of weathering aging coefficients on exposition conditions Legend: bold traces – not exposed to sun radiation, lightly traces – exposed to sun radiation.

Rubber samples, specially prepared without the required antioxidant protection show in a brief time the high aggressiveness of the environmental factors in different regions of Havana City.

Table 3. Components of rubber from national formulation number 471 without antioxidant protection

Ingredients	Parts by 100 parts of rubber (pphr)
Styrene butadiene – copolymer (E-1712)	50.00
Styrene butadiene - copolymer (E-1502)	20.00
Sorprene-293	30.00
Activator (Oxide of Zinc)	3.00
Stearic acid	2.00
Carbon Black (N-339)	60.00
Oil ICP-68	12.00
Vulcacit-CZ	1.10
TMTD	0.10
Vulcalent G	0.20
Sulphur	1.50
	$\sum$ 179.90

Table. 4. Initial physico – chemical properties of experimental rubber formulations

Samples	Modulus at 100 %, MPa	Modulus at 300%, MPa	Tensile strength, MPa	Elongation, %
1	2.8	9.93	19.32	480
2	2.69	11.16	20.39	480
3	3.08	11.4	19.16	480
4	2.71	12.36	21.24	_
5	2.17	-	_	-
Xm	2.7	11.2	20.0	480.0
σ	0.3	1.0	1.0	0.0

The rubber samples specially prepared without the required antioxidant protection are show on the following photographs (Figure 3). They expose in a short time of exposition to the atmospheric factors, the high aggressiveness present in two different regions of the city of Havana.

In the former photographs you may appreciate that all the changes on experimental rubbers occurred very fast. The incidence is higher in the station named "Flores" which is located only 20 meters from the northern coast in Havana city.

There are some other experiences related with the behavior of different rubber articles. Though these rubber articles were formerly protected with certain antioxidant levels, they also showed the aging symptoms in a short period of exploitation (months). These aging symptoms made them unuseful and as result they were rejected for use.

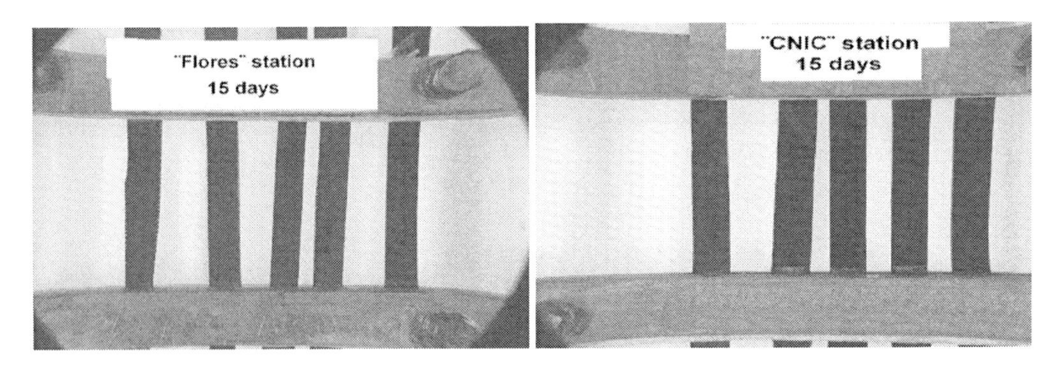

Figure 3. Rubber samples specially prepared without required antioxidant protection.

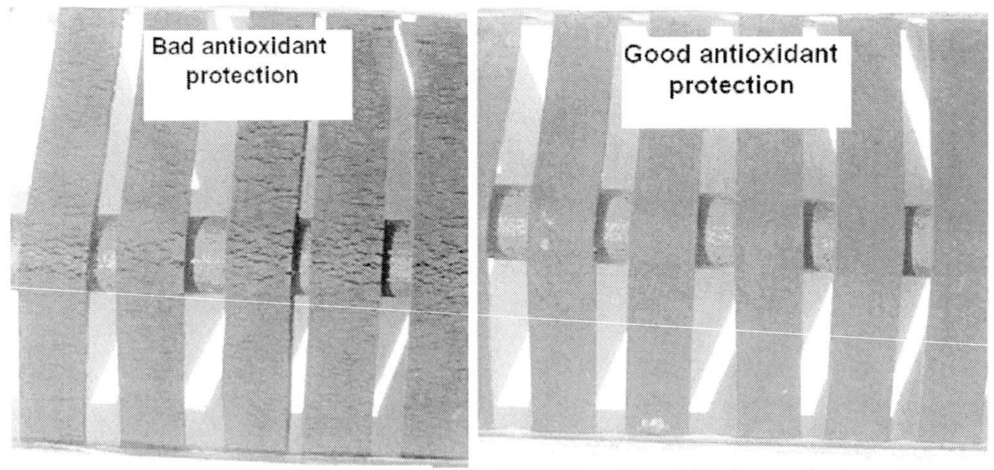

Figure 4. Different enhancement of rubber formulations with similar composition but with different antioxidant protection, during out door exposition and by strength.

The same has been observed in the case of rubber materials and articles under the action of the atmospheric factors from the Cuban environment, not only during their exploitation but also during the conservation. The following image shows what it has been formerly stated.

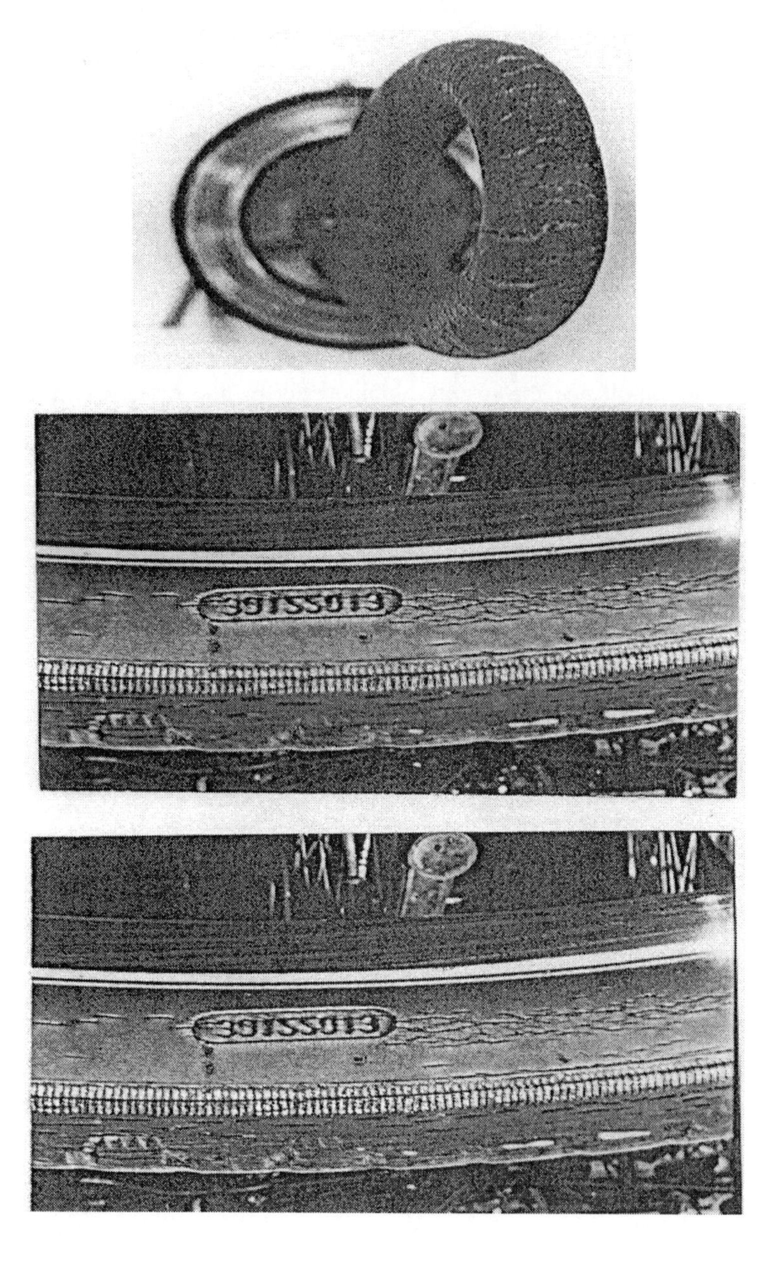

Figure 5.

CONCLUSION

The results show the high aggressiveness of the Cuban climate on polymeric materials and articles.

The research stations allow the evaluation not only of the resistance of the different materials but also the effectiveness of the different protecting systems whether you are using them or planning to use for the protection of the protection of the mentioned materials during their exploitation of and conservation in Cuba.

The high aggressiveness of the different Cuban regions is very similar to the conditions given by "Xenotex" climatic chambers which are the ones in use during quick evaluations.

Recommendations

The atmospheric research stations on materials from different kinds, may be use to evaluate the behavior of previously conceive and elaborated materials when they are under the action of the aggressive agents from the Cuban climate. These stations are useful for the selection of new protecting systems to prevent the aging process on different materials and also to confirm the validity of the results obtained during the researches recommended to carry out.

References

[1] F. Corvo, N. Betancourt, J.C. Díaz, C. Lariot, I. León, J. Pérez, O. Rodríguez, E. Bricuyet, F. Catalá, M. Castro, R. González, C. Echevarria, M. Lorente, M.E. Ladrón de Guevara, 2da variante del Mapa Regional de Agresividad Corrosiva de la Atmósfera de Cuba. Proceeding del PrimerTaller Internacional de Corrosión. CONACYT-CINVESTAV. Mérida. Yucatán. México. 23-28 marzo 1992.

[2] Razumovsky S.D, Podmasteryev V.V, Zaikov G.E Physico-Chemical Principles Governing the Ageing of Elastomers Under Action of Atmospheric Ozone *.Inter.J. Polymeric Mater.* 1990. Vol.13 pp 81-101.

[3] Rakovski S.K, and Cherneva D.R Sínthesis of Antioozonants for the preparation of Rubber. *Intern. J. Polymeric mater.*1990 Vol.14, pp 21-40.

[4] Popov A.A and Parfenov V.M. Efficiency of antiozonants under the conditions of static deformation of elastomers. *Intern. J.Polymeric Mater.* 1990, Vol. 13 ,pp 123-135.

[5] Ramirez Almoguea Jesús. Trabajo de Tesis para optar por el Grado Científico de Dr. En Ciencias Quimicas. Ciudad de la Habana. 1989.

[6] Ricard J.L Tesis presentada para optar por el grado científico de Dr. *En Ciencias Técnicas.*Ciudad Habana. 1994.

[7] Bolivar S., Reinosa O. Envejecimiento y estabilización de la goma para la banda de rodamiento de neumaticos de bicicletas de producción nacional. *Tesis de maestría.* Habana.1999.

[8] Reinosa O., Bolivar S., Ricard. J.L. Reforzamiento de la actividad antioxidante de una mezcla de aminas aromáticas empleadas en la protección de gomas al envejecimiento atmosférico. *Revista CNIC.Ciencias Quimicas* Vol. 29 pp 145-149 1998.

[9] Emanuel N.M. Some Problems of Chemical Kinetics in Polymer aging and Stabilization. *Polymer Engineering and Science* Vol.20, pp 662-67.

In: Polymer Research and Applications
Editors: Andrew J. Fusco and Henry W. Lewis

ISBN: 978-1-61209-029-0

Chapter 6

Peroxide Cross-Linking Polyethylene PEX-A: Migration of Antioxidants

T. L. Gorbunova, E. V. Kalugina and M. I. Gorilovsky
Polyplastic Group, Moscow, Russia

Abstract

Cross-linking polyethylene PEX-a were investigated. Cross-linking polyethylene pipes are exposed for boiling in hot water at 95oC. Analysis of migration antioxidants into hot water were investigated by Fourier-IR-spectroscopy method, UV-spectroscopy and liquid chromatography method (HPLC). Heat stability formulations PEX-a were investigated by DSC (method oxidation induction time OIT).

Keywords: antioxidants, polyethylene PEX-a, migration, hindered phosphite, hindered phenol, sulphur-containing phenol.

The issues of an additives migration are usually adverted to when polymeric materials are to be used in food-contact applications, e.g. as packaging material or in production of disposable dishware. The physicochemical examination of aqueous extracts of polymers is a traditional approach to the study of an antioxidants migration. The analysis of the domestic and foreign scientific and technical information referred to in earlier articles [1,2] has allowed specifying experimental conditions but has not given a concrete information on migration of antioxidant additives, which are of interest to us, from PEX-A into hot water. Therefore, an in-house program of experiments has been designed and developed by us.

Experimental

Test specimens: PEX-A pipe segments, produced by the standard peroxide (Engel) method, but with different formulations of stabilizer systems:

Formulation No 1: PEX-A pipe sample contained a synergetic mixture of antioxidants (AO): phenol antioxidant with melting point 54°C (PhAO 1) , sulphur-containing phenol antioxidant with melting point 70°C (SPhAO).

Formulatrion No 2: PEX-A pipe sample contained a liquid phenol antioxidant with melting point below 10°C (PhAO 2).

Formulation No 3: PEX-A pipe sample contained PhAO 2, which quantity is two times less than in Formula No 2.

Formulation No 4: PEX-A pipe sample contained a synergetic mixture of phenol antioxidant (PhAO 2) and hindered phosphite antioxidant (HPhAO).

The content of the above-mentioned antioxidant additives in PEX-A is given in Table 1.

Table 1. The content of the analyzed additives in PEX-A-formulations

Formulations	Additives content, %			
	PhAO 1	PhAO 2	SPhAO	HPhAO
Formulation No1	0.51	-	0.15	-
Formulation No2	-	0.66	-	-
Formulation No3	-	0.33	-	-
Formulation No4	-	0.51	-	0.15

Experiment

1. Thermal treatment of pipe segments in tap water at 95°C; exposition (days): 0 (before keeping process starts); 30; 60; 90; 180 and 360;
2. Thermal treatment of pipe segments (chips width 50 μm) in distilled water and ethanol at 60°C during 13 days.

Methods of Analysis

The method oxidation induction time (OIT) – ISO 113557-6 of the test specimens/samples was determined by differential scanning calorimetry method (DSC) with Pyris 6 DSK thermal analysis instrument from PerkinElmer, USA, at 210°C and oxygen flow of 20 ml/min.

1. The gross-evaluation of the presence of antioxidants in PEX-A pipes was made by Fourier-IR-spectroscopy method on Thermo-Nicolet Avatar 370 (single reflection ATR accessory designed)
2. The total content of phenol antioxidants was determined by UV-visible spectroscopy method on Thermo Electron Helios spectrometer with the preliminary extraction of the additives from the polymer (chips 50 μm) by the boiling chloroform for 16 hours.
3. The phosphite content (for Formulation No 4) analysis was done with the photocolorimetric method. The analysis of antioxidant content was carried out by the high performance liquid chromatography method (HPLC) on Thermo Electron

Spectra System with UV- visible photometric detector in a gradient mode. The chromatography column: Hypersil Gold ODS, 150x4mm; Thermo Electron Corp., particles' size 5 μm.

The mobile phase:
Eluent A: 75/25 methanol / water
Eluent B: 50/50 ethyl acetate / acetonitrile.
The gradient:

Time, minutes	Eluent A content, %	Eluent B content, %
0	90	10
20	0	100

The flow rate 1 ml/min, wave-length 270 nm, sample volume 20 μl.

4. The gel content was determined in accordance with ISO 10147.

Results and Discussion

The results of evaluation of the oxidative induction time of pipe samples before and after boiling in water at 95°C are given in Table 2. During experiment the pipe segments were washed by hot water, i.e. the experimental conditions chosen were much more severe than the real pipe service conditions with water running only inside the pipe. As the tests were carried out in hot running water, containing high levels of salts, ferrous salts in particular, the pipes' surface after experiment had yellow color because of the rust film. The OIT was assessed for cuts from the PEX-A pipe surface and from its inner part. Before the beginning of the experiment the pipes' surface was wiped with alcohol to maximally remove rust, but it was impossible to clean the surface completely (for it to become of white color). The inner layer (of white color) was drilled.

The data on changes of the OIT of PEX-A incorporating different antioxidant systems is given in Table 2.

Table 2. The OIT of PEX-A pipes during the process of boiling in hot water

Exposition, days	OIT (210°C) of samples after boiling in water at 95°C, min.							
	Formulation No 1		Formulation No 2		Formulation No 3		Formulation No 4	
	Surface layer	Inner layer	Surface layer	Inner layer	Surface layer	Inner layer	Surface layer	Inner layer
0	12,3	17,2	10,0	13,6	10,8	12,3	12,0	21,5
30	10,8	14,6	12,6	16,4	7,9	11,9	11,0	21,8
60	8,7	14,6	12,6	13,7	7,4	11,8	11,0	20,0
90	8,5	14,1	10,8	11,8	7,8	11,5	10,1	20,3
180	7,8	15,2	13,6	13,1	4,6	5,5	7,6	15,0
360	10,0	13,9	11,9	13,1	5,6	6,4	9,9	13,6

The received results showed that thermal stability of pipes produced from PEX-A prepared using the antioxidant, Formulations 2 and 3, but without synergists, was much lower if compared to that, when synergistic mixtures were used (Formulations 1 and 4). The use of phenol antioxidant only was not effective enough. Thermal stability of pipes made of PEX-A incorporating SPhAO or HPhAO was practically the same. It should also be taken into consideration that PhAO/synergist ratio in the formulas was not changed and so the effectiveness of the synergists (sulphur-containing and phosphite) was the same.

The results of our experiment fully correlate with data, provided by Dover Corporation. But in contrast to us its specialists have evaluated the OIT at 235°C in air. They have studied HDPE, stabilized by various additives after its exposure to water or 5% water solution of a bleach at 60°C during 5000 hours. It was demonstrated that the stabilizing system containing phosphite and PhAO was the most effective one.

The results of gel content in PEX-A pipes before and after its keeping in water are given in Table 3.

Table 3. The gel content changing in crosslinked PE containing different stabilizing systems after boiling in water at 95°C

Exposition, days	Formulation No1	Formulation No2	Formulation No3	Formulation No4
0	85,0%	86,5%	89,0%	88,29%
30	85,7%	84,7%	90,3%	90,5%
60	84,9%	84,4%	90,5%	90,1%
90	83,3%	85,5%	90,6%	90,3%
180	85,6%	82,7%	88,3%	88,8%
360	80,9%	82,1%	89,7%	87,3%

The received results showed that the gel content in the examined compositions has not practically changed.

Consumption of antioxidants was analyzed using several test methods:

Method 1. By optical density of absorption of carbonyl group in IR-spectra of PEX-A before and after boiling in hot water.

The –C=O absorption at 1740 cm^{-1} indicates the presence of antioxidants in PEX-A; this absorption band is absent in unstabilized PE. The characteristic absorption bands of PE are 2900 and 720cm^{-1} absorption bands (valence and deformation vibrations in CH_2).

As a case in point Figure 1 shows an original IR-spectrum of PEX-A (for Formulation No 1). IR spectra of PEX-A incorporating other antioxidant packages are absolutely identical.

A change of intensity (decrease) of an absorption band of antioxidant should indicate its consumption in the result of boiling in hot water.

Dependences of antioxidants consumption in Formulations No 1, No 2 and No 4 are presented in Figure 3. The content of phenol antioxidant in Formulations No 1 and No 2 is identical. The difference is that Formulation No 1 represents a synergetic mixture of two phenol antioxidants (PhAO+SPhAO) while Formulation No 2 contains only PhAO 2. The results received show that consumption of phenol antioxidants in Formulation No 1 is quicker than in Formulation No 2.

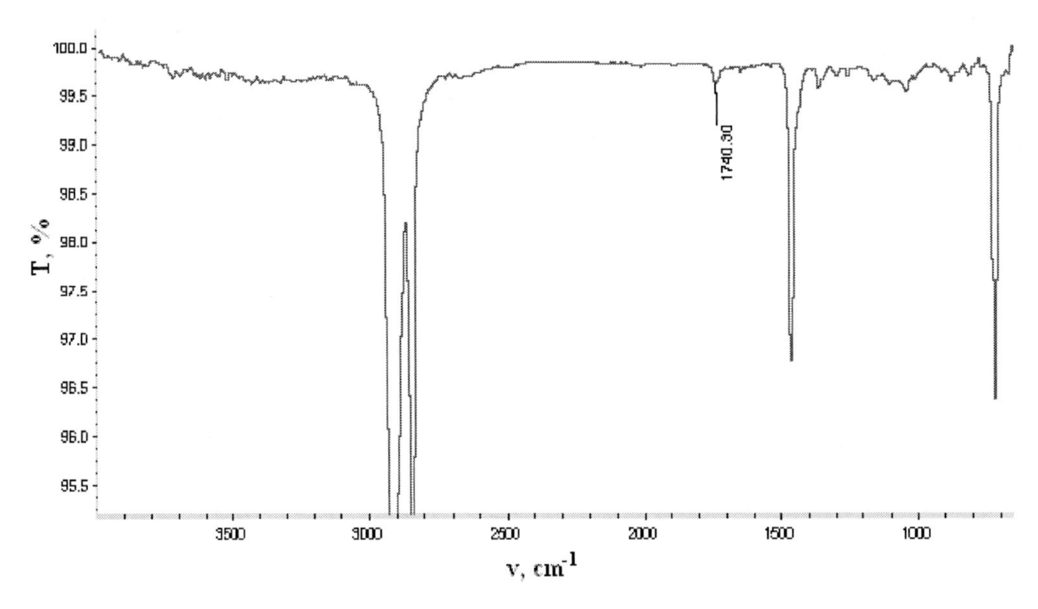

Figure 1. The original IR spectrum of stabilized PEX-A.

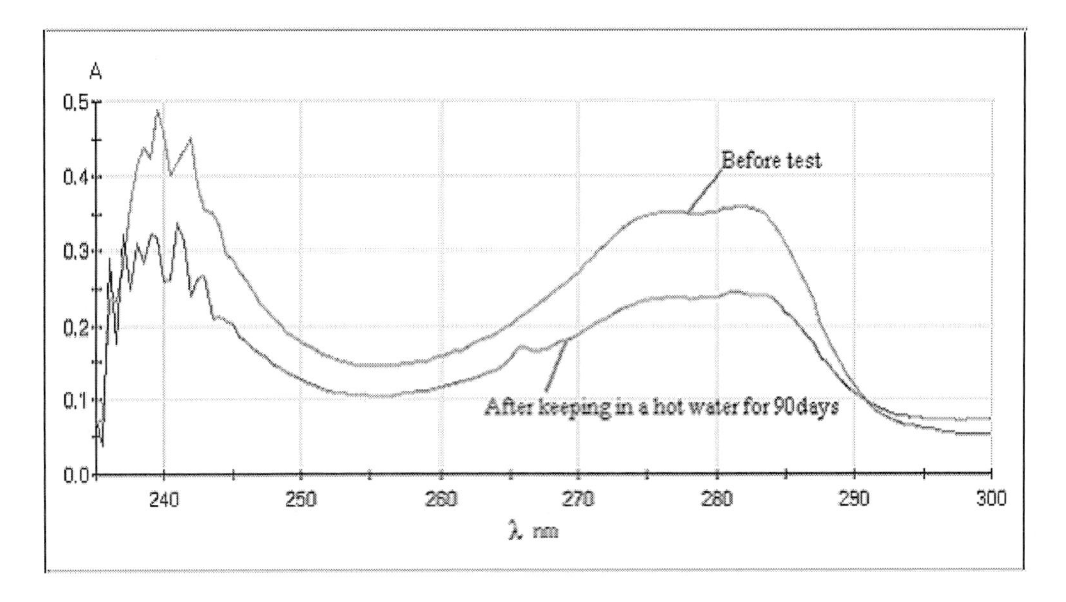

Figure 2. The original UV spectra of extracts from Formulation No 1 before and after boiling of pipes in water at 95°C. Solvent – chloroform.

The extract from Formulation No 4 after 60 days of keeping in water also shows a minor washing out of phenol antioxidant as judged from the change in the dynamics of its optical density. Aside from PhAO 2, this Formulation also contains HPhAO.

HPhAO content was analyzed separately (by phosphorus content) by the procedure, elaborated by Russian Research Institute of Plastics (NIIPM) ('Procedures for photometric determination of phosphites in polymers.). The results are presented in Table 5.

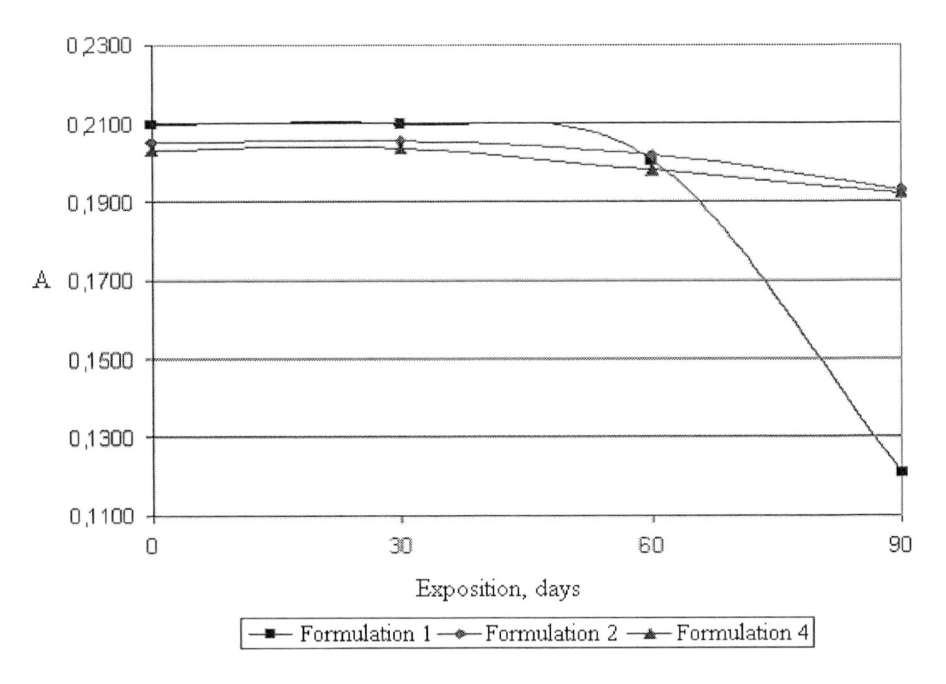

Figure 3. The change of absorption of extracts from samples after exposure of pipes in water at 95°C.

Table 5. HPhAO content in Formulation No 4

Exposition	Additive content, %
0 days	0,106
30 days	0,102
60 days	0,098
90 days	0,102
180 days	0,107
360 days	0,105

According to the results of phosphorus analysis, phosphite content in PEX-A pipes does not change during the entire time span under examination. But it is no secret that phosphorus is practically not washed out from polymers and even under the most severe conditions of the degradation it is being analyzed in coke [3].

Common phosphites, in their turn, are not hydrolytically stable and do easily break-down with formation of phenol and acid phosphite [4]:

$$(ArO\!-)_3P \xrightarrow{H_2O} \left[\begin{array}{c} HO\!\diagdown P(\!-\!OAr)_2 \\ \\ O\!=\!P(\!-\!OAr)_2 \\ \quad H\!\diagup \end{array}\right] + ArOH$$

It can be inferred that at boiling in water a special hydrolytically stable HPhAO selected by us was not virtually extracted in water. Still, according to the results of analysis carried out in RandD center of DOVER Corporation, even a pipe sample not subjected to thermal treatment in hot water in addition to the initial HPhAO also contains admixture of phosphate (0.0013%) and phenol (0.08%), i.e. phosphite conversion product. In a pipe sample after a year's boiling in water the phosphate content (0.08%) even exceeds that of phosphite (0.06%). The obtained results are quite reasonable and match the data of our analysis of antioxidant content, obtained by recalculation of data on phosphorus content.

Thereby, the conducted studies aimed at analysis of thermal stability of pipes made of PEX-A incorporating different antioxidant systems and of migration of antioxidants into hot water, showed that synergetic systems of antioxidants are practically identical in terms of their efficiency (Table 2).

A little lower migration-into-boiling-water capability of liquid antioxidants can be explained by their considerably more uniform distribution in PE powder.

Recommendations on express analysis of migration of low-polymeric substances from food-contact polymeric materials are given in EC Directives [5]. Following these guidelines we conducted a replicate experiment for Formulations No 1 and No 4.

Distilled water and ethanol were used as migration mediums (it excluded rust and other salts entering); test temperature - 60°C (as recommended); exposition period – 13 days. However, we decided to toughen experimental conditions. According to the guidelines, pipe segments of a determined weight and size were to be put into a liquid medium. But we prepared thin cuts (about 50 μm) and put them into water or ethanol, i.e. in our experiment the contact surface was significantly larger.

In case of the above-described experiment we analyzed antioxidant residues in finished products – PEX-A pipes - after their long-term exposure in hot water, and in the replicate experiment we analyzed the quantity of antioxidants migrating into liquid mediums. The analysis of antioxidant content was carried out by the high performance liquid chromatography method (Table 6).

Table 6. Migrated antioxidants content

Migration medium	Formulation No 1		Formulation No 4	
	PhAO	SPhAO	PhAO	HPhAO
Water	Traces (<0.01%)	Traces (<0.01%)	Not found	Not found
Ethanol	0.06%	Not found	0.06%	Not found

The experiment showed that:

(a) migration capability of phenol antioxidants into hot water is considerably lower than into ethanol;
(b) migration capability of phenol antioxidants, used in Formulations No 1 and No 4 is practically identical;
(c) migration capability of SPhAO is slightly higher than of HPhAO.

For comparison purposes we conducted a similar experiment (with exposure to distilled water at 60°C during 13 days) on pipes, made of PE 80 and PE 100. The following antioxidants were detected in water: phenol antioxidant – Irganox 1010; phosphite – Irgafos 168 and phosphate, its hydrolysate.

The received results fully correlate with factors influencing migration. Aside from temperature, specific surface area and time period of contact of polymer and experimental medium, these factors include:

(a) solubility of low-polymeric additives in polymer and liquid experimental medium. A maximum quick migration is taking place when additives are soluble in the experimental medium and insoluble in polymer. In case the substances are soluble in polymer but insoluble in the experimental medium migration is not taking place;
(b) moisture stability of polymer and additives. In case of low moisture stability of polymer even the moisture-stable additives can be washed out from material on attainment of maximum water absorption.

PE is characterized by a very low water absorption, e.g. HDPE water absorption index = 0.04%. All antioxidants tested by us are insoluble in water, but they are distributed in PE melt perfectly (the classical feature required of any antioxidant [7]). Furthermore, according to the notion of diffusion, in cross-linked structures such processes are hampered. That is why there is a low probability of the examined antioxidants migration from PEX-A - a conclusion, which is supported by the received experimental results.

REFERENCES

[1] E.V.Kalugina, T.L.Gorbunova Protection of health. The requirements of the control of polymer materials contacting with drinking water.Л. – *Russian plastics* 2007, №9, pp. 53-56.
[2] E.V.Kalugina, T.L.Gorbunova About migration of hazardous substances from polymer materials. – *Russian plastics* 2007, №8, pp. 52-55.
[3] N.A. Mukmeneva Phosphorylation like a way to increase the stability of polymers. *Proceedings of 8 All-Union school-seminar on hetero-organic compounds*.- M.: INEOS Academy of Sciences USSR. 1984 , 22 p.
[4] E.V.Kalugina, K.Z.Gumargalieva and G.E.Zaikov New Concept in Polymer Science: *Thermal Stability of Engineering Heterochain Thermoresistant Polymers*. VSP, Utrecht-Boston, 2004, 279 p.
[5] "Note for guidance for petitioners presenting an application for the safety assessment of a substance to be used food contact materials prior to its autorisation" 30.11.2004, Amsterdam.
[6] EAS on Paper Interim Report September 2001.
[7] *Plastic Additives Handbook*. Stabilizers, Processing Aids, Fillers, Reinforcements, Colorants for Thermoplastics. Ed. By R.Gachter and H.Muller. Hanser Publ., Munich Vienna, NY, 754 p.

In: Polymers Research and Applications
Editors: Andrew J. Fusco and Henry W. Lewis
ISBN: 978-1-61209-029-0

Chapter 7

PHYSICAL – CHEMICAL PROPERTIES OF NATURAL POLYMERS: POTENTIAL CARRIERS AND DELIVERY SYSTEMS OF BIOLOGICALLY ACTIVE SUBSTANCES FOR HUMAN APPLICATIONS

V. F. Uryash*[*1], A. V. Uryash[2], A. E. Gruzdeva[3], N. Yu. Kokurina[1], V. N. Larina[1], L. A. Faminskaya[1] and I. N. Kalashnikov[1]
[1]Research Institute of Chemistry, Nizhni Novgorod State University, Nizhni Novgorod, Russia
[2]Mount Sinai Medical Center, Department of Research, Miami Beach, Florida, USA
[3]Grande" Ltd., Nizhni Novgorod, Russia

ABSTRACT

Thermodynamic characteristics and physical-chemical properties of natural polymers (cellulose, starch, agar, chitin, pectin and inulin), their water mixtures and some biologically active substances extracted from vegetable substances using carbon dioxide in a supercritical state are reviewed. In addition, several aspects of practical application of thermodynamic characteristics of biologically active substances are demonstrated.

Keywords: polysaccharides, water, physical-chemical properties, biologically active substances, supercritical fluid extraction.

1. INTRODUCTION

Application of substances of natural origin for prophylactics of many diseases and increase of treatment efficiency is becoming a vital necessity. Actually all vegetables, fruits

[*] E-mail: ltch@ichem.unn.ru

and berries are the sources of vitamins which increase organism resistance to infections, stimulate nervous and muscular activity, help overcoming physical loads (strengthen vessel walls, affect the function of endocrine glands and hematosis, are necessary for proteic substance formation etc.). In addition to vitamins plants contain a lot of other components which are useful for human health, namely mineral salts (mainly those of alkali metals), organic acids, fatty and essential oils, tanning agents, phytoncides, cellulose, etc.

It is important to launch commercial production of complexes of biologically active substances (BAS) from edible vegetable raw materials and to utilize them in a concentrated form as biologically active additions to food with an aim to optimize nutrition of broad masses of population, mainly children.

However, when developing new technologies of vegetable raw processing and improving the existing ones the manufacturers face a number of problems which can be successfully solved using the results of the unique complex scientific investigations of thermodynamic properties of polymer-fluid systems. A physical-chemical analysis of such systems, BAS–water ones included, allows revealing crystallization of water surplus over its solubility in a natural polymer above the temperature of mixture vitrification and to plot diagrams of physical states of the aforementioned systems within a wide temperature range and in the entire range of component concentrations. Such diagrams allow determining temperature and concentration ranges of the system homogeneity and heterogeneity and help forecasting behavior of the systems for different fields of their practical application [1, 2]. The authors of the present review have been developing and analyzing physical state diagrams of complex muli-component polymer systems for over 35 years [3–5]. Diagram construction and analysis requires the data on vitrification temperature of BAS–water mixtures, water solubility in BAS (in other words concentration of a saturated water solution in BAS), as well as the melting temperature of the phase of water surplus over its solubility in BAS. Such data can be obtained when measuring heat capacity of BAS–water mixtures.

Based on $C_p^\circ=f(T)$ curves thermodynamic characteristics of substances (enthalpy, entropy and the Gibbs function) can be calculated. Thermodynamic characteristics of amino acids, proteins and polysaccharides should be determined both when studying the processes of their production and their transformations in living organisms, e.g., in enzymatic reactions or during conformational alterations of proteins and nucleic acids. This data can be found, e.g., in [6–14]. As a rule, biologically active substances are natural polymers (polysaccharides, proteins, etc.) with complex molecular and supramolecular structures. In this connection it is important to reveal regularities in the structure effect on their physical-chemical and, in particular, thermodynamic properties from both theoretical and practical points of view. Such generalizations can be found in [2, 3–5, 15].

Since the processes in organisms take place in a water environment it is important for their physiology and biochemistry to study water physical states in different BAS and to analyze the effect of water on temperatures of their physical transitions. Such data can be found, in particular, in [3–5, 16].

BAS transportation to a target organ in a human organism necessitates having polymers capable of performing transport functions. This role can be played, e.g., by cellulose and such its derivatives as starch, chitin, pectin, inulin, etc. It is necessary to know their thermodynamic characteristics to correctly select a BAS carrier.

The present review gives thermodynamic characteristics* and physical-chemical properties of the aforementioned natural polymers, their water mixtures** as well as some BAS extracted using carbon dioxide in a supercritical state. In addition, some aspects of practical application of thermodynamic BAS characteristics are demonstrated.[1]

2. Thermodynamic Characteristics of Polysaccharides and Their Water Mixtures

2.1.1. Thermodynamics and Physical-Chemical Analysis of Cellulose and Enthalpy of Its Interaction with Water

There are literary data on heat capacity of cotton and wood cellulose [17–24] studied within the range of 80–450 K using vacuum adiabatic and differential scanning calorimeters. When analyzing different samples of cotton cellulose some researchers [18–21] revealed a strong dependence of its low-temperature heat capacity on the conditions of treatment and thermal processing of polysaccharide (Figure 1). This result is quite unexpected since the polymers of one and the same chemical structure in crystalline and amorphous states do not demonstrate, as a rule, significant differences in thermodynamic characteristics at the temperatures lower than the vitrification temperature [25, 26]. A similar picture is observed for low-molecular substances as well [27–29]. This very regularity was found by the authors of [17] when measuring low-temperature heat capacity of wood cellulose of different crystallinity degree within the range of 80–400 K with an error of <0.5 % using a vacuum adiabatic calorimeter. The averaged values of C_p° and thermodynamic functions are listed in Table 1. To calculate thermodynamic functions the authors of [17] extrapolated cellulose heat capacity to 0 K according to the Kelly-Parks-Hoffman method [30] taking C_p° of the cellulose nitrate containing 11.9 mass % of nitrogen measured from 4 K as a reference [4, 31].

As far as cellulose is concerned, several physical relaxation-type transitions in different temperature ranges were registered using different techniques [18, 23, 32–43]. It is attributed to heterogeneity of its structure, namely to the fact that in cellulose there exist highly-ordered and amorphous microregions [44–46]. Thus in [32, 33, 36, 39, 43] it is found using dilatometric, thermo-mechanical and dielectric techniques that the ordered cellulose regions devitrificate at 490—500 K. Vitrification of amorphous cellulose regions at 450 K was revealed using the dynamic mechanical technique [40].

---

[1] *In the present review thermodynamic characteristics of polysaccharides are calculated per a molar mass of the repeated group of their macromolecules, g/mol: cellulose, amylose and inulin – 162.14; amylopectin – 145.13; agarose – 306.27; chitin – 203.19; chitosan – 161.16;

**Molar mass of polysaccharide mixtures with water (Mmix) was calculated with the formula: Mmix=M1X1+M2X2, where M1 and M2 are the molar masses of a repeated polymer group and H2O, respectively; X1 and X2 are the molar fractions of the polymer and H2O in the mixtures, respectively.

Table 1. Heat capacity and thermodynamic functions of wood cellulose samples with crystallinity induces 0 (I), 0.4 (II) и 0.65 (III) [17]

T, K	C°_{p}, J/mol K			$S^{\circ}(T)-S^{\circ}(0)$, J/mol·K			$H^{\circ}(T)-H^{\circ}(0)$, kJ/mol			$-[G^{\circ}(T)-G^{\circ}(0)]$, kJ/mol		
	I	II	III	I	II	III	I	II	III	I	II	III
10	3.375	3.316	3.266	1.901	1.867	1.838	0.0125	0.0123	0.0121	0.0065	0.0064	0.0063
50	33.33	32.42	32.07	24.14	23.88	23.30	0.7225	0.7093	0.6970	0.4843	0.4759	0.4679
100	67.13	66.15	64.53	57.83	56.98	55.66	3.248	3.191	3.123	2.534	2.506	2.443
150	100.6	97.40	96.50	91.37	89.98	88.24	7.441	7.315	7.199	6.264	6.181	6.037
200	134.3	132.8	131.0	124.9	123.0	121.0	13.31	13.10	12.92	11.67	11.50	11.27
250	168.1	166.0	164.4	158.5	156.2	153.8	20.87	20.57	20.31	18.76	18.48	18.14
300	205.1	201.0	198.7	192.3	189.5	186.7	30.16	29.72	29.36	27.53	27.12	26.65
350	254.1	246.0	238.4	227.7	223.8	220.4	41.67	40.89	40.30	38.02	37.45	36.82
400	303.0	290.7	278.0	265.0	259.7	254.7	55.66	54.34	53.20	50.33	49.53	48.70

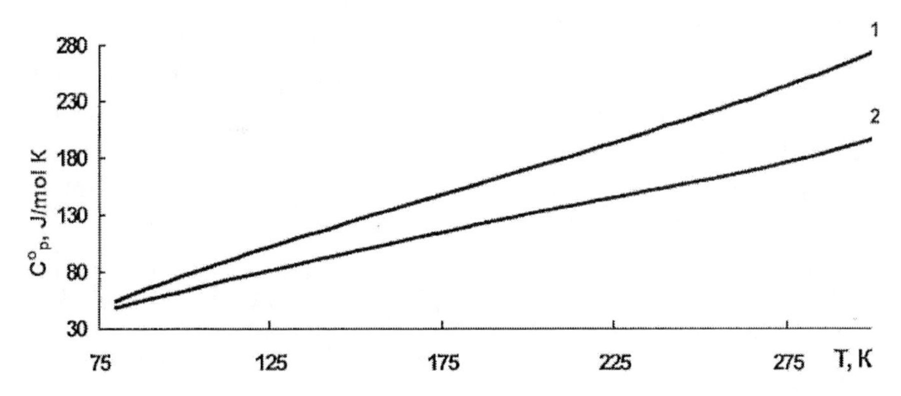

Figure 1. Temperature dependence of heat capacity of cellulose I dehydrated by inclusion from Russian (1) and foreign (2) raw [20].

At the temperatures lower than T_g several secondary low-scale physical transitions were registered in cellulose. For example, in [36, 41] the transition at 180—200 K is related to the methyl group stability and polarization of the primary hydroxyl in the electric fieldAs far as the transition at 273–310 K in cellulose and its derivatives are concerned, the researchers [32, 38, 42] attribute them to libration of pyranose rings around a glucoside bond or to conformational alteration of an “armchair-bath” type at the glycopyranose group level.

The papers [47—49] give the results of thermal conductivity studies within the 80–330 K range and DTA of cotton microcrystalline cellulose (MCC) in the range of 80–550 K. It is apparent from Figure 2 that MCC heat capacity monotonously increases within the range of 80—280 K. Then a minor endothermic anomaly appears on the $C_p^{\circ}=f(T)$ curve, its temperature determined based on the $C_p^{\circ}/T=f(T)$ plot being 284 K.

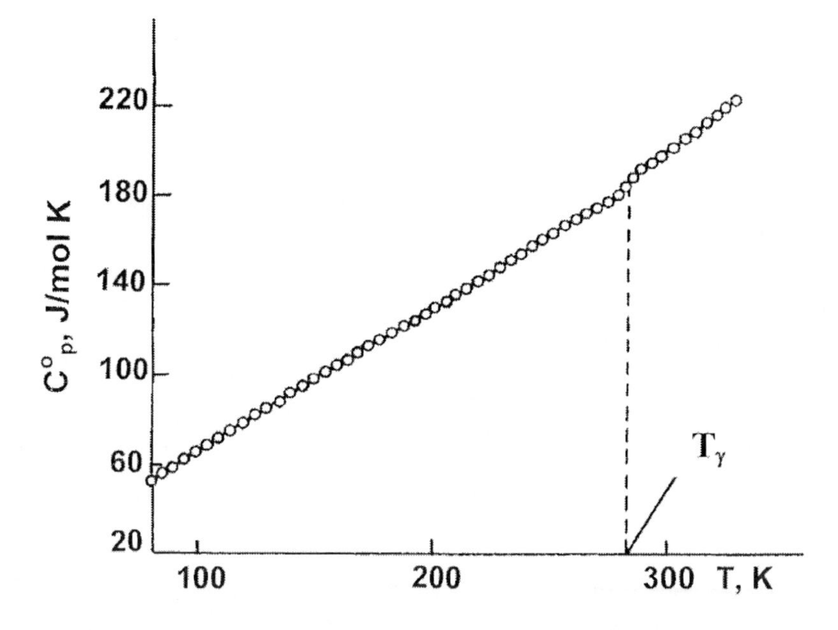

Figure 2. Mole heat capacity of cotton microcrystalline cellulose [47-49].

The thermogram of dehydrated MCC (Figure 3) demonstrates three relaxation transitions. One of them which takes place at 291 K is analogous to the process observed on the $C_p^{\circ}=f(T)$ curve. It can be classified as γ–transition due to oscillations of the side groups in a cellulose macromolecule. The second one observed at 343 K is attributed to β–transition attributed to excitation of pyranose ring oscillations around glucoside bonds. The last transition whose mean temperature is 403.5 K is associated with vitrification of disordered MCC microregions.

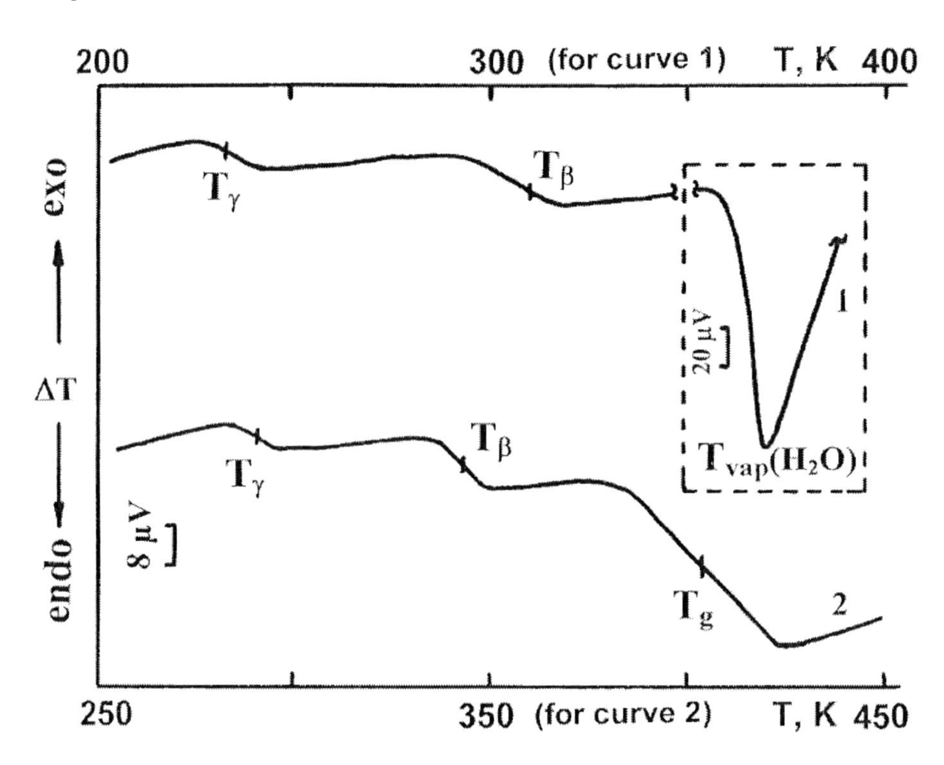

Figure 3. Thermograms of air-dry (1) and dehydrated (2) microcrystalline cellulose [47-49].

Burning and formation enthalpies are thermodynamic characteristics which are sensitive to chemical and physical structures of substances. The validity of this statement as referred to cellulose is confirmed by a number of publications [24, 45, 50–53]. Standard burning enthalpy of cellulose samples of different origin obtained using different treatment techniques varies from –2844 kJ/mol up to –2781 kJ/mol. The exothermic effect of polysaccharide burning reaction should decrease with an increase in the cellulose degree of order since it is necessary to spend some energy to destruct highly-ordered polymer microregions. On the contrary, the absolute value standard enthalpy of cellulose formation will increase with an increase in its degree of order. This very tendency was revealed when determining $\Delta_c H^{o}$ и $\Delta_f H^{o}$ of MCC and wood cellulose with different crystallinity indices [48, 49] (Table 2). An analog dependence was observed in [50].

Natural and engineering processes with cellulose participation most often take place in water environment. Thus it is important to analyze the process of its interaction with water and, particularly, the accompanying energy effect as a function of cellulose degree of order.

Table 2. Standard energy and enthalpy of cellulose sample burning and formation at 298.15 K (kJ/mol) [49]

Sample*	$-\Delta_c U^o$	$-\Delta_c H^o$	$-\Delta_f H^o$
(I)	2843.5±4.4	2843.5±4.4	946.7±4.4
(II)	2837.0±2.9	2837.0±2.9	953.2±2.9
(III)	2834.3±2.4	2834.3±2.4	956.0±2.4
MCC	2802.3±0.8	2802.3±0.8	988.0±0.8

*The symbols for wood cellulose samples (I, II, III) are similar to those in Table 1.

It is known from literature [54–64] that the process of interaction of the given polysaccharide with water is accompanied by an exothermic effect. In addition cellulose hydrophily and, as a result, the enthalpy of cellulose interaction (mixing) with water ($\Delta_{mix}H$) change depending of the degree of order [57–62, 64]. The weaker the bonds between cellulose macromolecules, the larger the number of OH-groups accessible for hydration with water molecules and the larger the absolute value of ($\Delta_{mix}H$). The enthalpies of interaction with water point, first of all, to the ratio of amorphous and ordered regions in cellulose [60, 61].

When interpreting the enthalpy of the process of cellulose interaction with water it is necessary to take into account not only the contribution of newly-formed hydrogen bonds during component mixture, but also the contribution attributed to the transition of the vitreous polysaccharide fraction into a highly-elastic state under the effect of the low-molecular component below the vitrification temperature [60, 61, 65–69]. It is so-called excess enthalpy of the vitreous state and it is negative. The excess enthalpy can be high enough for polar polymers with a high vitrification temperature, e.g. as is the case with cellulose. The calculations performed in [57, 58] showed that in the solution of an entirely amorphous vitreous cellulose the excess enthalpy of its transition from a vitreous to a highly-elastic state is –12.6±2 kJ/mol. With a temperature increase its absolute value decreases and above the vitrification temperature polymer interaction with a fluid component depends on the character and the value of intermolecular interaction of the components [60, 68].

The authors of [57, 58] got a linear correlations between the values of $\Delta_{mix}H$ of cellulose with water. They carried out the experiment using an adiabatic calorimeter [13, 70, 71] at 298 K with an error of 1.5 % and found out that for entirely crystalline cellulose $\Delta_{mix}H$ with water is close to zero, whereas for entirely amorphous cellulose it makes –27.1 kJ/mol. Equation (1) relating the enthalpy of cellulose mixing with water and the degree of polysaccharide crystallinity (X, %) was proposed:

$$\Delta_{mix}H \text{ (kJ/mol)} = -27.1 \bullet (1 - X/100) \quad (1)$$

It confirms the opinion [44, 56, 63] that water cannot penetrate cellulose crystalline regions. Thus, an inverse problem, namely, evaluation of the crystallinity degree based on thermochemical data, can be solved. For example, using the enthalpy of cotton MCC interaction with water ($\Delta_{mix}H$= –3.5 kJ/mol of polymer) obtained in [48, 49] the degree of MCC crystallinity can be estimated using Eq. (1), this value being 86 %.

2.1.2. Thermodynamics and Physical-Chemical Analysis of Chitin/Chitosan and Enthalpy of Their Interaction with Water

Heat capacity of chitin from carposome of young species of hiratake (*Pleurotus ostreatus, Fr.*) within the range of 12–320 K as well as that of chitin and chitosan from crab shell within the range of 80—320 K was measured using a vacuum adiabatic calorimeter [4, 5, 72–74] (Figure 4). Thermodynamic characteristics (enthalpy, entropy, the Gibbs function) of fungic chitin in the temperature range from T→0 K up to 320 K were calculated (Table 3). It is apparent from Figure 4 that heat capacity of chitin extracted from different sources monotonously increases in the analyzed temperature range. The fact that C_p° of fungic and crab chitin are different shows that chitin properties, in particular, its structure, drastically differ depending of the raw it is extracted from and the technique of its extraction and purification. This fact is also paid attention to in [75—83] and is verified by the results of the analysis of $C_p^{\circ}=f(T)$ for crab chitin and chitosan samples subject to acid hydrolysis [84]. A elevation on the $C_p^{\circ}=f(T)$ curves for chitin and chitosan at 290 K (Figure 4) seems to be attributed to the beginning of the relaxation transition in the samples under study.

DTA of fungic chitin in the range of 80–550 K (Figure 5) was considered in [4, 5, 72, 73]. The thermogram of dehydrated chitin from young fungi (Figure 5, curve 1) has three endothermic relaxation transitions. The first transition in the range of 300–330 K with the mean temperature (T_{β}) 316 K is a secondary β-transition. Two other transitions, namely 350-390 K (T_{g1} = 376 K) and 405-470 K (T_{g2} = 437 K) are attributed to devitrification. At that, the temperature of the first transition (T_{g1}) is lower than that of the second one (T_{g2}).

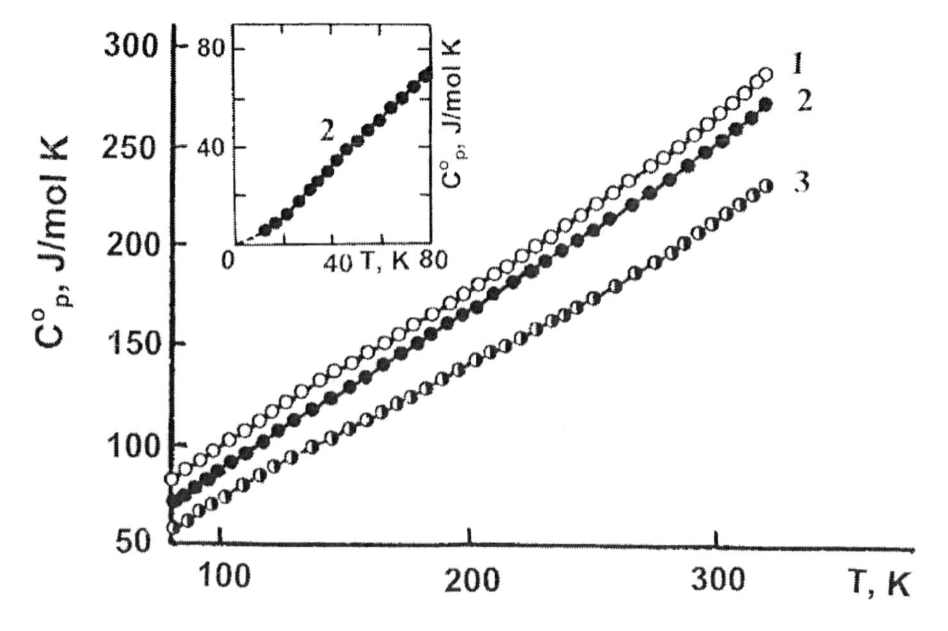

Figure 4. Mole heat capacity of chitin (1) and chitosan (3) from crab shells, chitin from carposome of young hiratake (*Pleurotus ostreatus, Fr.*) species (2) [4, 5, 72-74].

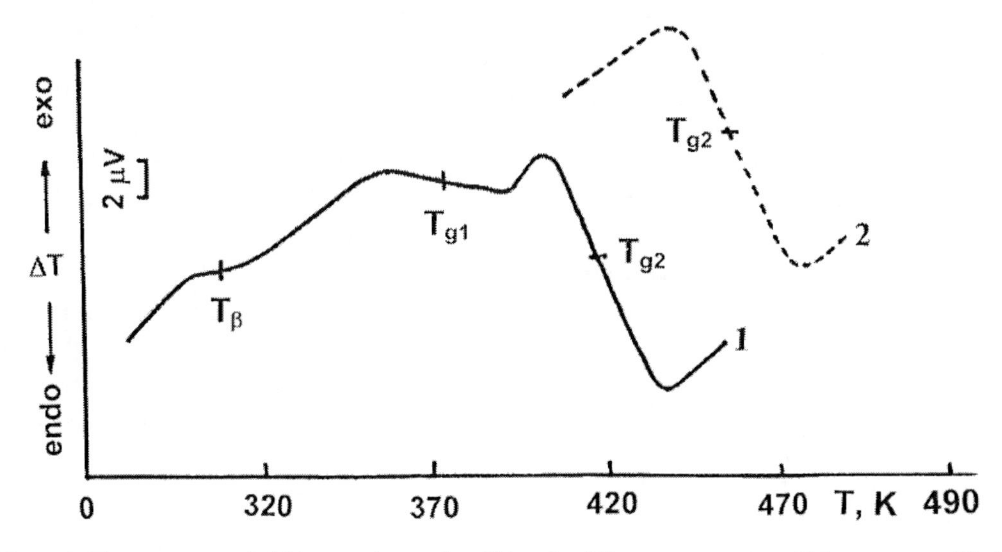

Figure 5. Thermograms of chitin samples made of hiratake (*Pleurotus ostreatus, Fr.*) carposome (2): 1-young species; 2 – old species [4, 5, 72-74].

Table 3. Averaged heat capacities and thermodynamic functions of chitin extracted from carposome of young hiratake (*Pleurotus ostreatus, Fr.*) species [4, 5, 72]

T, K	C_p°,J/mol·K	H°(T)–H°(0),kJ/mol	S°(T)–S°(0),J/mol·K	–[G°(T)–G°(0)],kJ/mol
10	3.441	0.0124	1.890	0.006469
20	10.51	0.0797	6.275	0.04580
40	31.20	0.4929	19.77	0.2979
60	51.45	1.323	36.33	0.8564
80	69.30	2.535	53.63	1.756
100	85.89	4.088	70.89	3.001
120	102.5	5.973	88.03	4.590
140	119.2	8.188	105.1	6.525
160	135.7	10.74	122.1	8.793
180	151.9	13.61	139.0	11.40
200	167.7	16.81	155.8	14.35
220	183.7	20.32	172.5	17.64
240	200.0	24.16	189.2	21.25
260	217.1	28.33	205.9	25.20
280	235.1	32.85	222.6	29.49
298.15	251.9	37.27	237.9	33.67
320	273.2	43.01	256.5	39.07

The temperatures of these transition for fungic chitin are lower than those for the crab one [4, 5, 84] that points to a lower degree of order of fungic chitin. A similar character of relaxation is observed in cellulose nitrates and acetates [2-4, 85-112]. Some physical transitions in chitosan were discovered by I.F. Kaimin'sh and his research team [113, 114].

The authors of [4, 5, 72, 73] attribute β-transition in chitin, as well as in cellulose, its acetates and nitrates, to libration of pyranose rings around a glucoside bond, whereas two temperature devitrification intervals are associated with inhomogeneous structure of chitin which (as cellulose) consists of amorphous and highly-ordered regions [75–84]. At that, a macromolecular chitin spiral is a more labile structure than that of cellulose [79, 80]. The results of the investigation of enthalpy of crab chitin and cellulose interactions with water and other solvents [115] and chitin sorption properties [82, 83] point to a lower degree of order in crab chitin as compared to cellulose. The authors of [116] determined interaction enthalpy of chitin extracted from hiratake carposome with water at 303 K (the ratio of component masses being 1:40). $\Delta_{mix}H = -18.9$ kJ/mol of polymer.

A peculiarity of fungic chitin, namely a dependence of temperatures of its relaxation transitions (T_{β},T_{g1},T_{g2}) on the fungi age should be paid attention to. The temperature of chitin from old fungi T_{β} and T_{g1} increased by ~10 K; T_{g2}, by 30–40 K (Figure 5, curve 2). It means that the chitin structure becomes more ordered with the fungi aging. After second devitrification decomposition is observed on fungic chitin thermograms. It helium atmosphere it is accompanied by energy absorption and is two-staged. For crab chitin endothermic peaks correspond to 485 K and 556.5 K [84].

The paper [84] considers the effect of acid hydrolysis on physical-chemical properties of chitin and chitosan extracted from crab shells. Heat capacity in the range of 80–330 K was measured with the error of 0.3 % and DTA of highly-ordered, so-called "crystalline" chitin (CT) and chitosan (CTS), as well as CT treated with the 80 % solution of phosphoric acid and CTS treated with 1 M of acetic acid solution, was performed in the range of 80–600 K. Such treatment lowered CT molecular mass from $1.4*10^6$ up to $1.0*10^6$ D; that of CTS, from $8.3*10^5$ up to $6.0*10^5$ D. A degree of order of polymers decreased as well that is confirmed by the results obtained in [84] (Figures 6, 7; Table 5).

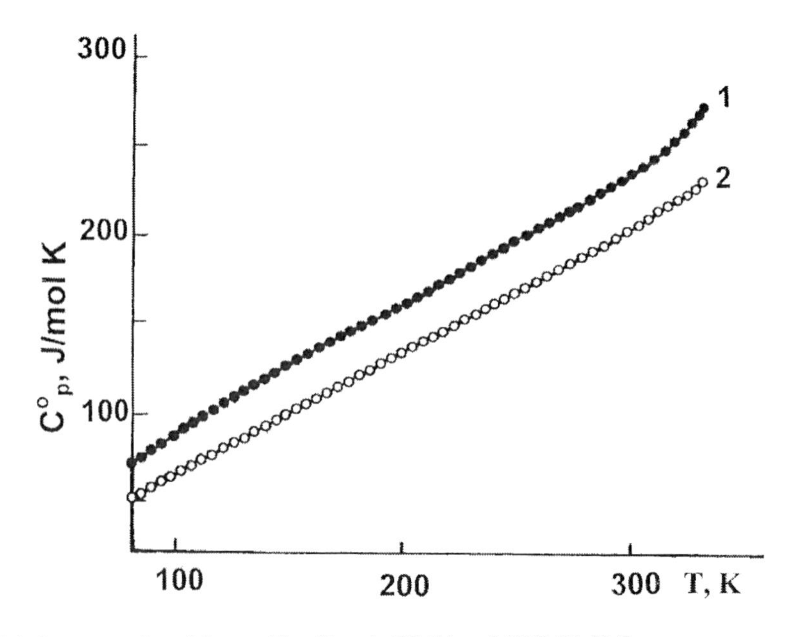

Figure 6. Mole heat capacity of "crystallized" crab CT (1) and CTS (2) [84].

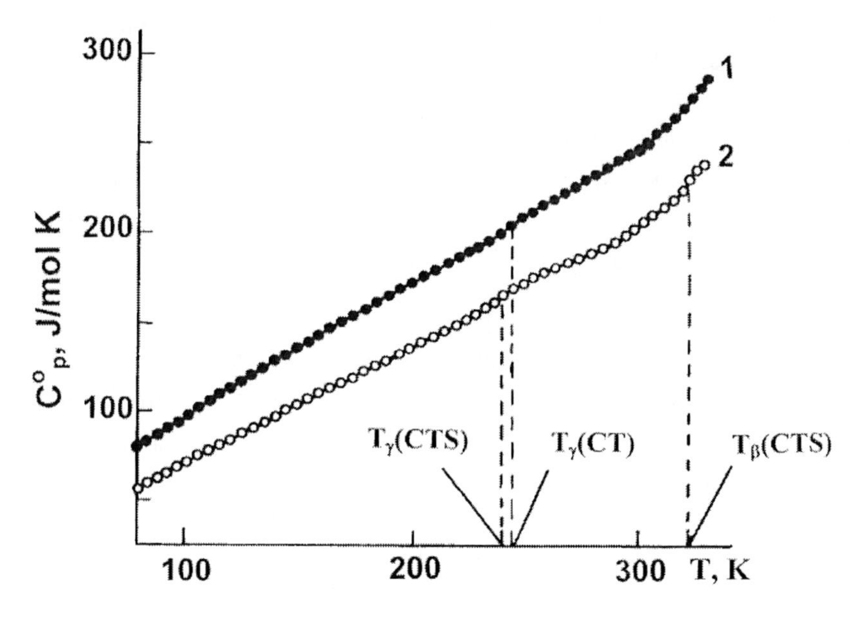

Figure 7. Mole heat capacity of hydrolyzed crab CT (1) and CTS (2) [84].

Table 4. Temperatures of physical transitions in analyzed samples from crab shell chitin (CT) and chitosan (CTS) determined calorimetrically and using DTA technique [84]

Sample	Adiabatic calorimetry		DTA		
	T_γ, K	T_β, K	T_β, K	T_{g1}, K	T_{g2}, K
CT «cryst.»	–	The beginning at 310 K	365	428	In the range of decomposition
CT hydrol.	246	The beginning at 310 K	338	403	432
CTS «cryst.»	–	The beginning at 310 K	326	361	417.5
CTS hydrol.	240	322	320	347	413

Table 5. Standard energies and enthalpies of burning for the samples from crab shell chitin (CT) and chitosan (CTS) as well as from fungic chitin at 298.15 K (kJ/mol) [4, 5, 72, 73, 84]

Sample	$-\Delta_c U^o$	$-\Delta_c H^o$	$-\Delta_f H^o$
CT «cryst.»	3833.8±5.4	3334.4±5.4	1171.5±5.4
CT hydrol.	3877.3±2.0	3377.9±2.0	1128.0±2.0
CT fungic	4005.8±16.7	4006.4±16.7	999.6±16.7
CTS «cryst.»	3162.7±4.2	3163.3±4.2	769.8±4.2

It is apparent from Figures 6 and 7 that molecular heat capacity of chitins is higher than that of chitosans within the entire temperature range that corresponds to a larger molar mass of the repeated CT group as compared to CTS. Comparison of heat capacities of "crystalline" and hydrolyzed CT shows that it is higher for the first candidate as compared to the second one. It points to the fact that the structure of "crystalline" CT is more ordered as compared to the hydrolyzed one.

Endothermic anomalies appear on $C_p^\circ=f(T)$ curves of crab hydrolyzed chitin and chitosan in the range of 230–250 K. They can be attributed to excitation of oscillations of minor side groups in CT and CTS macromolecules as well as to the presence of oligomer molecules formed during treatment of the samples [117-119]. They can be classified as γ-transition [2–5, 85-112]. Amine and acetamide groups act as such groups. The aforementioned transition is absent in "crystalline" CT and CTS. It can point to the fact that in the process of hydrolysis CT and CTS structure becomes less ordered.

In addition, one more relaxation transition is observed on the $C_p^\circ=f(T)$ curve for hydrolyzed crab CTS in the range of 315–325 K (Figure 7, curve 2). This relaxation process can be considered as β-transition [117-119] due to libration of pyranose rings around a glucoside bond. The beginning of this transition at 310 K is registered both in hydrolyzed CT (Figure 7, curve 1) and "crystalline" CT and CTS (Figure 6; Table 4).

Decomposition of CT (in helium atmosphere) is accompanied by an endothermic effect and is two-staged (T_{dest1}=485 K, T_{dest2}=556.5 K). When air-dry CT contains 4.0 mass % of water, the temperatures of physical transitions decrease (T_β=310 K, T_{g1}=344 K). It points to plasticizing effect of water on CT. At 406.5 K there appears an endothermic peak of sorption water evaporation.

Hydrolized anhydrous crab CT demonstrates the same transitions as those in chitin (Table 4), they appearing at lower temperatures (T_β=320 K, T_{g1}=347 K, T_{g2}=413 K). It can point to a lower degree of order of hydrolyzed CTS structure as compared to regenerated CT. As far as CTS is concerned, there is only one difference: its decomposition (in helium atmosphere) is exothermic. The first peak is observed at T_{dest1}=510 K. The energy released in the form of heat in the temperature range of the second peak is so high that it yields sample heating from 540 up to 590 K followed by its cooling up to 557 K. The water (8.0 mass%) contained in the air-dry CTS also exerts plasticizing action thus lowering the temperature of physical transitions (T_β=311 K, T_{g1}=288 K).

It is apparent from Table 4 that the temperature of β-transition increases when turning from hydrolyzed CTS to "crystalline" CT. Their T_{g1} and T_{g2} change in the same manner, the temperature of the second vitrification for "crystalline" CT falling into the temperature range of its decomposition. Based on the results obtained we can conclude that the structure of "crystalline" CT is more ordered as compared to regenerated CT. The same concerns CTS. Thus, chitin ordering increases in the process of its deacetylation and acid hydrolysis.

The papers [4, 5, 72, 73, 84] give the results of determining standard enthalpies of burning ($\Delta_c H^\circ$(solid, 298.15), kJ/mol) and formation ($\Delta_f H^\circ$(solid, 298.15 kJ/mol) of chitin and chitosan from crab shell as well as fungic chitin (Table 5). It is apparent from Table 5 that the exothermal effect of chitin burning reaction decreases with an increase in the degree of ordering of the given polysaccharide, i.e. during transition of fungic chitin to hydrolyzed and "crystalline" crab chitin. An analog dependence is traced for cellulose samples (Table 2). The authors of [120] obtained close values of enthalpy of crab shell chitin burning and formation.

The enthalpy of chitin and chitosan interaction with water ($\Delta_{mix}H$) at 298.15 K (the ratio of component masses being 1:40) (Table 6) was determined in [116, 121] using Tian-Calvet type differential microcalorimeter [122].

Table 6. Results of determining enthalpy of chitin (CT), chitosan (CTS) and microcrystalline cellulose (MCC) with water at 298.15 K [47-49, 117, 122]

Sample	CT «cryst.»	CT hydrol.	CTS «cryst.»	CTS hydrol.	CT fungic	MCC
$-\Delta_{mix}H$, kJ/mol	3.1_5	11.7	11.0	16.0	18.9	3.5

It is apparent from Table 6 that there is a correlation between polysaccharide degree of order, $\Delta_{mix}H$ of their mixtures with water and water solubility in polysaccharides. It is noted in [60, 61] that the electron-donor ability in the cellulose-chitin-chitosan series increases and exothermicity of interaction with acid solutions increases in the same direction (it decreasing in case of interaction with base solutions). It is attributed to the fact that the OH-group near the second carbon atom in the cellulose pyranose ring is substituted by the acetamide (chitin) and amine (chitosan) groups. The authors of [116, 121] did not reveal such a unique dependence. Actually, the enthalpy of interaction with water in case of a transition from "crystalline" chitin to "crystalline" chitosan and from hydrolyzed chitin to hydrolyzed chitosan becomes negative. However, the comparison with cellulose shows that ordering of the polysaccharide structure is not less important. It is likely that ordering of "crystalline" chitin is comparable with MCC ordering and fungic chitin is less ordered from among the studied chitins.

2.1.3. Thermodynamics and Physical-Chemical Analysis of Agarose and Agar

The authors of [4, 73, 123] measured heat capacity of dehydrated samples of agarose and agar in the range of 12—320 K using a vacuum adiabatic calorimeter (Figures 8, 9). As qualitative and quantitative compositions of the repeated agarose group was found in [124], heat capacity and thermodynamic functions are calculated per a molar mass of its repeated group (306.27 g/mol) (Table 7). On the contrary, agar is a mixture of agarose and agaropectin, the structure of the latter being ambiguous. Thus, the authors of [4, 73, 123] give specific heat capacity of agar. It is apparent from Figures 8 and 9 that heat capacities of agarose and agar monotonously increase within the range of 12–235 K. Then an anomalous heat capacity increase is observed on $C_p^\circ = f(T)$ curves in the range of 235–260 K. One more analog transition is observed in the 312–322 K range. The transition in this range is analogous to β-transition observed for fungic chitin, crab chitin, chitisan and cellulose nitrates and acetates [3-5, 31, 32, 38, 72-74, 84-114, 116, 121]. It is explained by libration of pyranose rings near a glucoside bond. The relaxation process at lower temperatures can be classified as γ-transition [117-119]. It seems to be attributed to excitation of side group oscillations in the macromolecules of agarose, the latter forming agar as well. Thus, similar relaxation transitions in the low temperature range are observed for agarose and agar. It points to the fact that as far as agar is concerned, they are not associated with the presence of agaropectin.

Table 7. Averages mole heat capacity and thermodynamic functions of agarose [4, 73, 124]

T, K	C°_{p},J/mol·K	H°(T)–H°(0),kJ/mol	S°(T)–S°(0),J/mol·K	–[G°(T)–G°(0)],kJ/mol
10	5.493	0.02008	3.065	0.01057
20	16.11	0.1249	9.913	0.07334
40	44.81	0.7336	29.86	0.4609
60	73.01	1.914	53.41	1.290
80	99.54	3.641	78.07	2.604
100	124.8	5.886	103.0	4.415
120	148.4	8.624	127.9	6.724
140	171.5	11.83	152.5	9.529
160	194.8	15.49	177.0	12.82
180	217.8	19.61	201.2	16.61
200	240.7	24.20	225.4	20.87
220	262.5	29.24	249.4	25.62
240	288.8	34.72	273.2	30.85
260	309.0	40.67	297.0	36.55
280	332.0	47.09	320.8	42.73
298.15	354.3	53.31	342.3	48.74
300	356.8	53.97	344.5	49.38
325	390.5	63.34	374.5	58.36

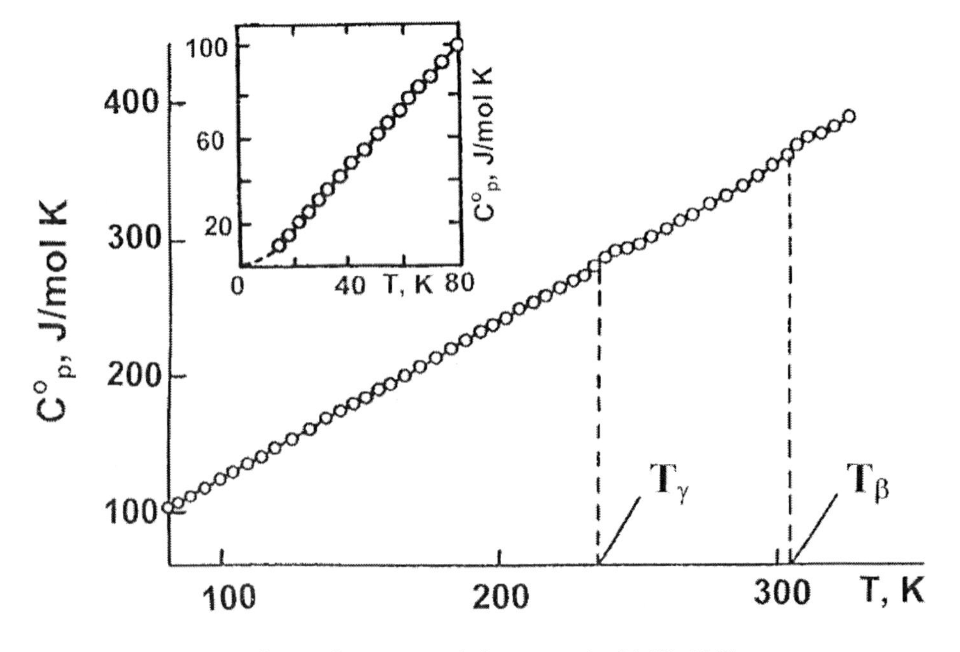

Figure 8. Temperature dependence of agarose mole heat capacity [4, 73, 124].

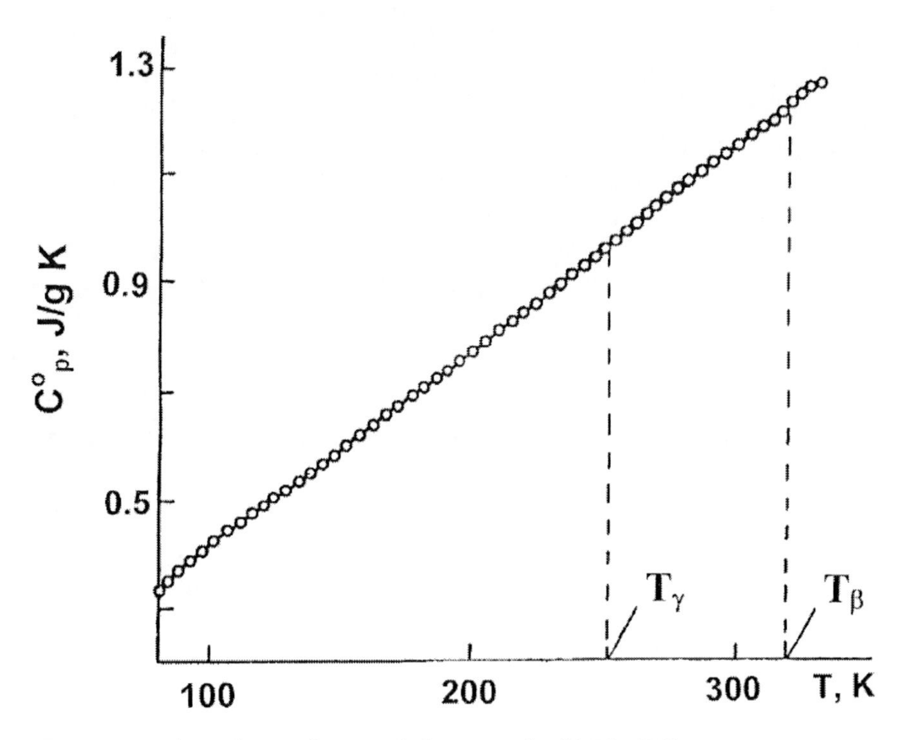

Figure 9. Temperature dependence of agar mole heat capacity [4, 73, 124].

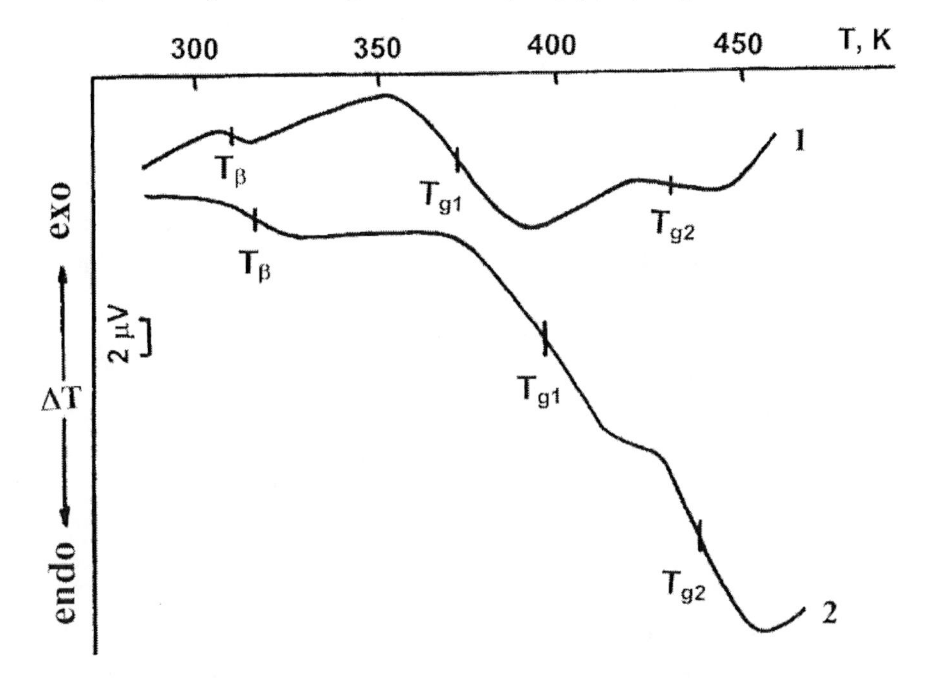

Figure 10. Thermograms of dehydrated agar (1) and agarose (2) [4, 73, 124].

The performed DTA of agar and agarose samples (Figure 10) [4, 73, 123] showed that there are three relaxation transitions in both substances which (by analogy with cellulose and its derivatives) can be classified as β-transition and two vitrifications T_{g1} and T_{g2}. As is the case with cellulose derivatives, two temperature ranges of vitrification can be explained by the fact that there are regions with different ordering in agarose and agar. As it is apparent from Figure 10, vitrification temperatures of agarose are higher than those of agar. It points to a less ordered structure of agar as compared to agarose that can be attributed to the presence of agaropectin in agar. The process of agar and agarose destruction also indicates it. In agarose it was one-staged with the exothermic peak extremum at 521 K. Destruction in agar was three-staged, the extremums being observed at 490, 506 and 556 K.

Standard enthalpies of agarose burning and formation in a solid state at 298.15 K were determined [4, 73, 123]. $\Delta_c H^o$(solid, 298.15) = −5735.4±13.3 kJ/mol, $\Delta_f H^o$(solid, 298.15) = −1559.2±13.3 kJ/mol.

2.1.4. Thermodynamics and Physical-Chemical Analysis of Amylose, Amylopectin and Starch

The experimental heat capacity values of amylose (AML), amylopectin (AMP) and potato starch in the range of 8-320 K [4, 73, 125] are shown in Figures 11–13. It is apparent from Figures 11–13 that heat capacity of AML and starch monotonously increases in the range of 80–290 K, that of AMP, in the range from T→0 up to 240 K. A sharper increase on the heat capacity curve for AML at 300 K (Figure 11) and starch at 290 K (Figure 13, curve 1) is attributed to β-transitions. This relaxation refers to the motions of polymer chain sections which are smaller than a segment [117–119]. In polysaccharides it can be attributed to excitation of pyranose ring oscillations near glucoside bonds [3-5, 31, 32, 38, 72-74, 84-114, 116, 121, 123]. The authors of [125] observed abnormal AML behavior at the temperatures lower than 80 K (Figure 11). Two endothermic anomalies with the mean temperatures 56.5 K and 65 K, determined based on $C_p^\circ/T=f(T)$ plots appeared on the $C_p^\circ=f(T)$ curve in the range of 45–75 K. These transitions are reversible since they were reproduced during several repeated heatings. The aforesaid anomalies seem to be of relaxation character and are likely to be attributed to oscillations of the CH_2OH–group of the 6th carbon atom of a pyranose cycle. It is in agreement with the absence of such transitions in AMP (Figure 12) since a glycoside bond in it is established in the 6th atom of carbon. In starch (Figure 13, curve 1) these transitions appear on the $C_p^\circ=f(T)$ curve at the temperatures 56 K and 61.5 K, it actually coinciding with AML. However, they are somewhat smoothed. Analog anomalies were observed, e.g., on the heat capacity curve for polym-ethylphenylsiloxane [126] and carboxylane dendrimer of the first generation [127].

The authors of [125] registered on the AMP heat capacity curve (Figure 12) two relaxation transitions at 245.0 and 297.5 K which they marked as $T_{\gamma 1}$ and $T_{\gamma 2}$, respectively. Based on the classification proposed in [117–119] γ-relaxation is attributed to internal rotation in the side groups. If there are several side groups in a polymer chain, several γ-transitions can be observed. The mentioned transitions are absent on the $C_p^\circ=f(T)$ curve for AML (Figure 11). It can be explained by peculiarities of the branched AMP structure.

Based on the data of the authors of [125] no γ-transitions of amylopectin were revealed on the $C_p^\circ=f(T)$ curve of starch.

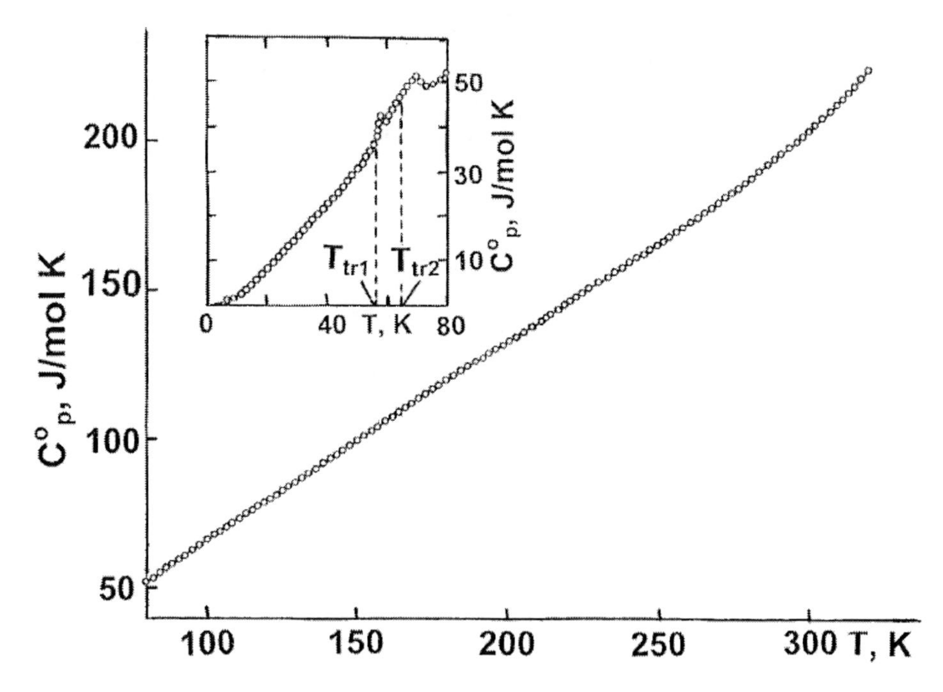

Figure 11. Temperature dependence of amylose mole heat capacity [4, 73, 126].

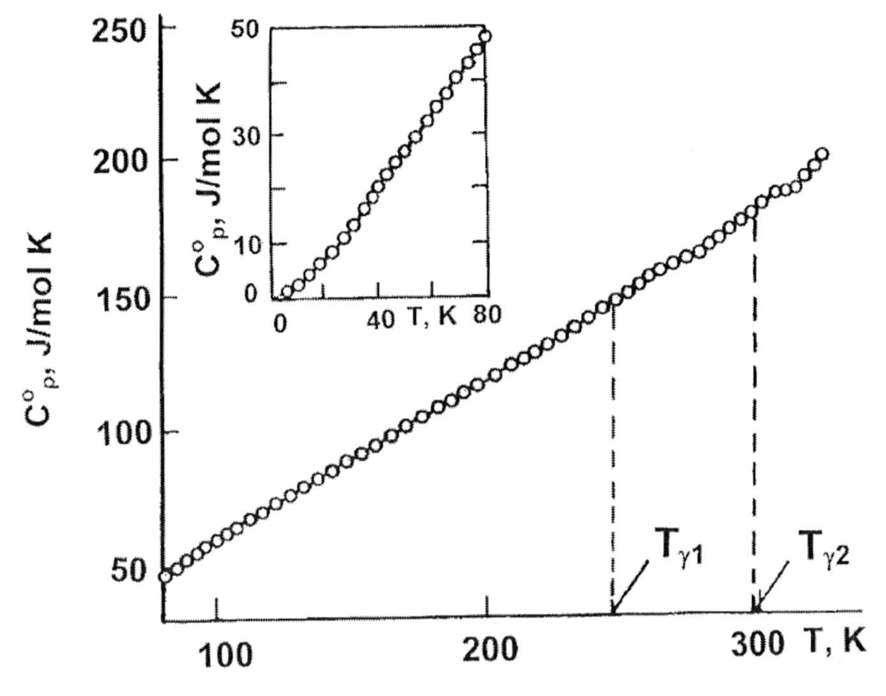

Figure 12. Temperature dependence of amylopectin mole heat capacity [4, 73, 126].

It seems to be attributed to the fact that starch, which consists of AML (33.5 mass %) and AMP (66.5 mass %), is not a mechanical mixture of these polysaccharides. There is interference between AML and AMP and γ-transitions of the latter are suppressed.

Table 8. Averaged heat capacity and thermodynamic functions of amylose (1), amylopectin (2) and potato starch (3) [4, 73, 125]

T, K	C_p°,J/mol·K			$H^o(T)–H^o(0)$,kJ/mol			$S^o(T)–S^o(0)$,J/mol·K			$–[G^o(T)–G^o(0)]$,kJ/mol		
	1	2	3	1	2	3	1	2	3	1	2	3
10	2.380	1.415	1.038	0.008356	0.003735	0.002403	1.258	0.5031	0.3201	0.004225	0.001295	0.000798
20	7.900	6.458	6.694	0.05770	0.04079	0.03832	4.463	2.872	2.588	0.03155	0.01665	0.01343
40	23.02	19.95	20.68	0.3622	0.3011	0.3083	14.41	11.36	11.39	0.2144	0.1531	0.1472
60	42.24	33.12	36.55	0.9955	0.8317	0.8748	26.98	21.93	22.65	0.6233	0.4844	0.4841
80	51.97	47.38	47.98	1.965	1.635	1.716	40.87	33.39	34.68	1.305	1.036	1.058
100	65.67	60.81	61.22	3.142	2.720	2.809	53.94	45.44	46.81	2.253	1.824	1.872
120	79.21	73.00	73.33	4.591	4.061	4.156	67.12	57.63	59.06	3.463	2.855	2.931
140	93.07	84.61	85.81	6.315	5.636	5.746	80.38	69.74	71.29	4.938	4.129	4.235
160	106.4	96.51	98.53	8.310	7.447	7.590	93.68	81.82	83.58	6.678	5.644	5.783
180	119.8	108.3	110.7	10.57	9.496	9.683	107.0	93.87	95.89	8.685	7.401	7.578
200	133.0	119.6	122.7	13.10	11.78	12.02	120.3	105.9	108.2	10.96	9.399	9.619
220	146.3	131.2	134.8	15.89	14.28	14.59	133.6	117.8	120.4	13.50	11.64	11.90
240	159.9	144.3	147.1	18.95	17.03	17.41	146.9	129.8	132.7	16.30	14.11	14.44
260	173.8	157.8	160.0	22.29	20.05	20.48	160.2	141.9	145.0	19.37	16.83	17.21
280	188.2	168.6	173.9	25.90	23.33	23.82	173.6	154.0	157.3	22.71	19.79	20.24
298.15	202.5	180.0	187.9	29.45	26.48	27.10	185.9	164.9	168.7	25.97	22.68	23.19
320	224.9	196.5	205.8	34.10	30.59	31.40	200.9	178.2	182.6	30.20	26.43	27.03

Table 9. Standard energy and enthalpy of burning and formation of amylose, amylopectin and starch at 298.15 K (kJ/mol) [125]

Sample	$-\Delta_c U^o$	$-\Delta_c H^o$	$-\Delta_f H^o$
AML	2822.9±3.8	2822.9±3.8	967.3±3.8
AMP	2521.5±1.7	2522.1±1.7	1125.3±1.7
Potato starch	2615.7±1.6	2616.1±1.6	1075.3±1.6

Nonadditive dependence of starch heat capacity on C_p° of AML and AMP can serve as an evidence [126]. A minor (1–2.5 %) negative deviation of starch C_p° from additive straight lines in the range of 100–250 K was observed. Such dependence points to interaction of AML and AMP macromolecules in starch. Averaged heat conductivity values and thermodynamic functions for AML, AMP and starch are listed in Table 8 [4, 125].

Standard enthalpies of burning [$\Delta_c H^o$(solid, 298.15), kJ/mol] for AML, AMP and starch were determined in [125]. Based on the obtained values standard enthalpies of their formation in a solid state [$\Delta_f H^o$(solid, 298.15), kJ/mol] were calculated (Table 9). The authors of [125] noted that the value of $\Delta_f H^o$ (starch, solid) is additively made from $\Delta_f H^o$ (AML, solid) and $\Delta_f H^o$ (AMP, solid), their concentration in starch being taken into account. It shows that intermolecular interaction in starch does not affect the enthalpy of its formation.

2.1.5. Thermodynamics and Physical-Chemical Analysis of Pectin

Pectins are very useful to people [124]. They regulate the work of the bowels and have detoxication properties, e.g., in case of mercury poisoning. As it is noted in [128], pectin has a wide spectrum of effects on the digestive system. It is capable of sorbing and leading out of the organism cholesterol, bile acids, urea, bilirubin and exert bactericidal action on pathogens of infectious diseases of human gastrointestinal tract, the bowels microflora being preserved. An important property of pectin is its ability to form a gel due to intermolecular association [129].

The authors of [4, 130] measured heat capacity of apple pectins, Classic brand, manufactured by HerbstreithandFox (FRG) in the range of 6–330 K and performed their DTA in the range of 80–450 K. The samples of AS-401 (with etherification degree of 63–65%) and AU-202 (with etherification degree of 68–76%) pectins were analyzed. The molar mass of the repeated AS-401 group was 185.10 g/mol; AU-202, 186.22 g/mol. The experimental results of measuring pectin heat capacity are demonstrated in Figure 14. It is apparent from Figure 14a that heat capacity of pectins with different degrees of etherification monotonously increases in the temperature range under analysis. Heat capacity of the pectin with a greater etherification degree (AU-202) is slightly higher (curve 1). However this difference is negligible and in the range of 250–265 K heat capacities of the samples coincide. A heat capacity increase at 300 K can be attributed to β-transition which, according to DTA data, begins at T_β=320 K (AS-401) and T_β=332 K (AU-202). A reversible endothermic anomaly was observed on the heat capacity curve for AS-401 pectin sample in the range of 45–60 K.

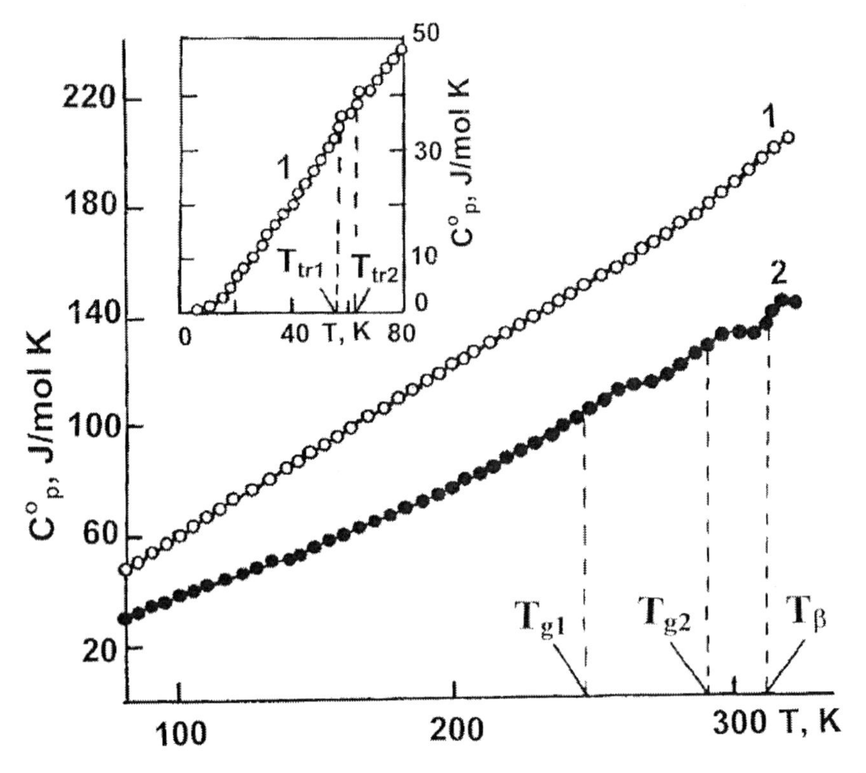

Figure 13. Temperature dependence of mole heat capacity of potato starch-water mixture samples containing , mass% of H_2O: 1–0; 2–12.2 [4, 73, 126].

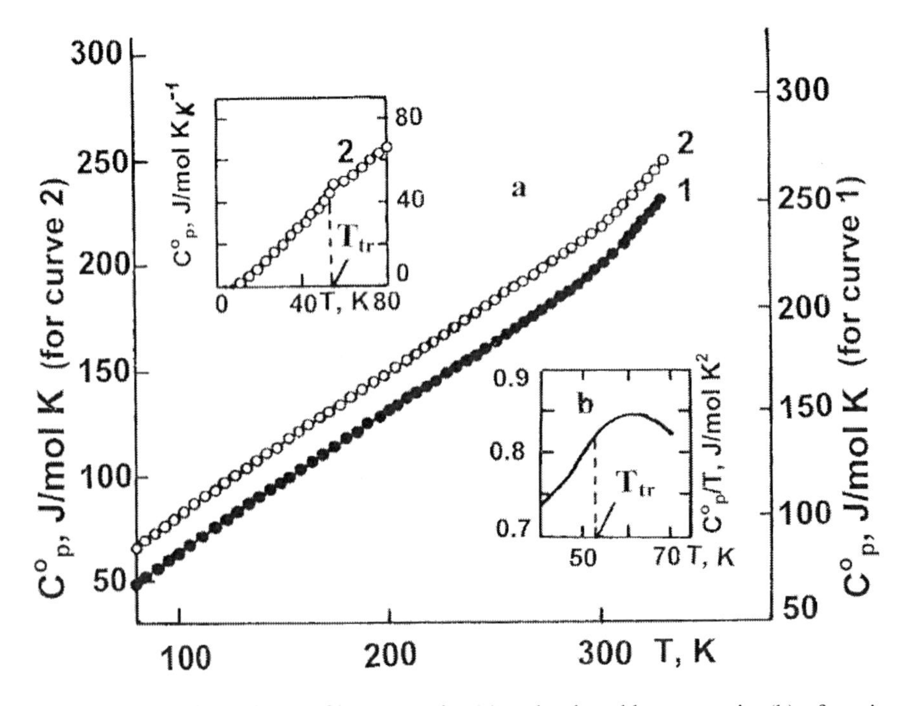

Figure 14. Temperature dependence of heat capacity (a) and reduced heat capacity (b) of pectins: 1 – AU-202; 2 – AS-401 [4, 131].

The mean transition temperature (T_{tr}), determined based on the $C_p^{\circ}/T=f(T)$ plot is 53 K. It seems to be attributed to excitation of methyl group oscillations near the 6th carbon atom of the pyranose ring. It should be noted that such anomaly was observed in amylose and starch [125]. In addition to β-transition two more vitrification temperature ranges appeared on the pectin thermograms [4, 130]. Their mean temperatures are T_{g1}=332.5 K and T_{g2}=404.5 K for AS-401, and T_{g1}=354 K and T_{g2}=399 K for AU-202. Other polysaccharides behave in a similar manner [3-5, 31, 32, 38, 72-74, 84-114, 116, 121, 123, 125]. It is attributed to complex molecular and supramolecular structures of polysaccharides. They consist of highly-ordered and disordered microregions. Pectine destruction is accompanied by absorption of energy in the form of heat. The endothermic peak is observed at 420 K.

The thermodynamic functions of AS-401 and AU-202 pectins in the range from T→0 up to 320 K [4, 130] are listed in Table 10. The enthalpies of burning [$\Delta_c H^o$(solid, 298.15), kJ/mol] of AU-202 and AS-401 pectins were obtained [4, 130]. Based on the obtained values the enthalpy of their formation in a solid state was calculated [$\Delta_f H^o$(solid, 298.15), kJ/mol] (Table 11).

Table 10. Averaged heat capacity and thermodynamic functions of AS-401 (1) and AU-202 (2) pectins [4, 131]

T, K	C_p°,J/mol·K		$H^o(T)-H^o(0)$,kJ/mol		$S^o(T)-S^o(0)$,J/mol·K		$-[G^o(T)-G^o(0)]$,kJ/mol	
	1	2	1	2	1	2	1	2
5	0.1883	0.1993	0.000234	0.000248	0.06266	0.06636	0.000079	0.000084
10	1.620	1.713	0.003811	0.004032	0.5067	0.5361	0.001256	0.001330
20	9.550	10.07	0.05581	0.05892	3.797	4.010	0.02014	0.02129
40	29.31	30.74	0.4440	0.4670	16.46	17.33	0.2146	0.2262
60	50.27	50.12	1.237	1.283	32.25	33.61	0.6979	0.7331
80	65.40	67.84	2.390	2.472	48.73	50.59	1.508	1.575
100	80.33	82.87	3.850	3.983	64.96	67.38	2.646	2.755
120	94.96	97.30	5.606	5.785	80.92	83.77	4.105	4.268
140	108.7	111.3	7.645	7.871	96.61	99.83	5.881	6.104
160	122.4	125.4	9.952	10.24	112.0	115.6	7.967	8.259
180	135.9	138.3	12.53	12.87	127.2	131.1	10.36	10.73
200	149.0	150.5	15.38	15.76	142.2	146.3	13.05	13.50
220	162.4	163.5	18.49	18.90	157.0	161.3	16.05	16.58
240	175.8	176.2	21.88	22.30	171.7	176.0	19.33	19.95
260	189.7	189.0	25.53	25.95	186.3	190.6	22.91	23.62
280	202.9	202.6	29.46	29.86	200.9	205.1	26.78	27.58
298.15	214.7	216.4	33.24	33.66	214.0	218.3	30.55	31.42
320	235.3	238.5	38.15	38.62	229.8	234.3	35.40	36.36
330	248.5	250.0	40.56	41.07	237.3	241.8	37.73	38.74

Table 11. Standard energy and enthalpy of burning and formation of Classic AS-401 and AU-202 pectins at 298.15 K (kJ/mol) [4, 131]

Sample	$-\Delta_c U^o$	$-\Delta_c H^o$	$-\Delta_f H^o$
AS-401	2965.2±3.3	2963.5±3.3	975.7±3.3
AU-202	3046.0±4.1	3044.3±4.1	949.3±4.1

The obtained results show that the standard enthalpy becomes more negative with an increase in the degree of pectin substitution, the situation with the formation enthalpy being the opposite.

2.1.6. Thermodynamics and Physical-Chemical Analysis of Inulin and Its Mixtures with Water

Contrary to the aforesaid polysaccharides inulin should be ascribed to oligomers. Its polymerization degree is low and makes about 35 monosaccharide residua. The molecular mass of inulin is 5000. A macromolecule of this polysaccharide is made of D-fructose residua [124, 129, 131]. The structure of its macromolecule can be represented as GF_n, where G are the glycosyl enzymes, F_n are the fructosyl enzymes and n is the number of bound fructosyl enzymes. Inulin plays a role of a food reserve. It was discovered in tubers of many Compositae, e.g. dahlia, chicory, dandelion and artichoke. Inulin substitutes starch in these plants. Inulin in dahlia and artichoke tubers makes 50 % of the live tissue mass. As inulin easily dissolves in hot water and precipitates during cooling, it can be extracted by water and purified using recrystallization. Inulin methylation points to the fact that D-fructose residua are bound with 2→1 linkages and are in a furanose form. β-configuration of glycosidic linkages was determined based on low specific rotation of inulin. The absence of regenerative properties and the presence of minor amounts of D-fructose in hydrolyzates are attributed to the fact that the regenerative end of inulin macromolecule is substituted by a glycoside–glycoside group of saccharose type. Inulin is easily hydrolyzed with acids and thus this reaction is used to produce fructose from inulin-containing raw. A specific enzyme, namely inulasa, which also hydrolizes inulin yielding fructose formation, is present in mold fungi and yeast.

A favorable effect of inulin is used to treat such illnesses as diabetes mellitus, dysbacteriosis, viral hepatitis, diseases of gastrointestinal tract as well as alimentary obesity and skin diseases [132]. Inulin ability to selectively stimulate bifido– and lactobacteria, they being representatives of normal bowels microflora, allows calling this substance as prebiotic.

Heat capacity of inulin extracted from chicory roots (ICN Pharmaceutical, Inc.) was measured and the temperatures of its physical transitions were determined in [4, 133] (Figure 15). For the sake of comparison heat capacity of refined wood cellulose "Tyrecell" (98 % of α–cellulose) with a zero crystallinity index is given in the same figure (curve 2) [17].

It is apparent from Figure 15 that inulin heat capacity monotonously increases in the range of 80–190 K. Then a minor endothermic anomaly appears on the $C_p^\circ = f(T)$ curve, its temperature (199 K) being determined based on the $C_p^\circ/T = f(T)$ plot [4, 133]. The aforementioned anomaly on the $C_p^\circ = f(T)$ inulin curve can be attributed to γ-relaxation which is likely to take place due to oscillations of side groups in a polysaccharide macromolecule. A heat capacity rise on the $C_p^\circ = f(T)$ inulin curve, which begins at 300 K, is attributed to β-transition in it. The comparison of chicory inulin heat capacity with that of amorphous wood cellulose [17] shows that in inulin they are close below the transition temperature (Figure 15, curves 1, 2).

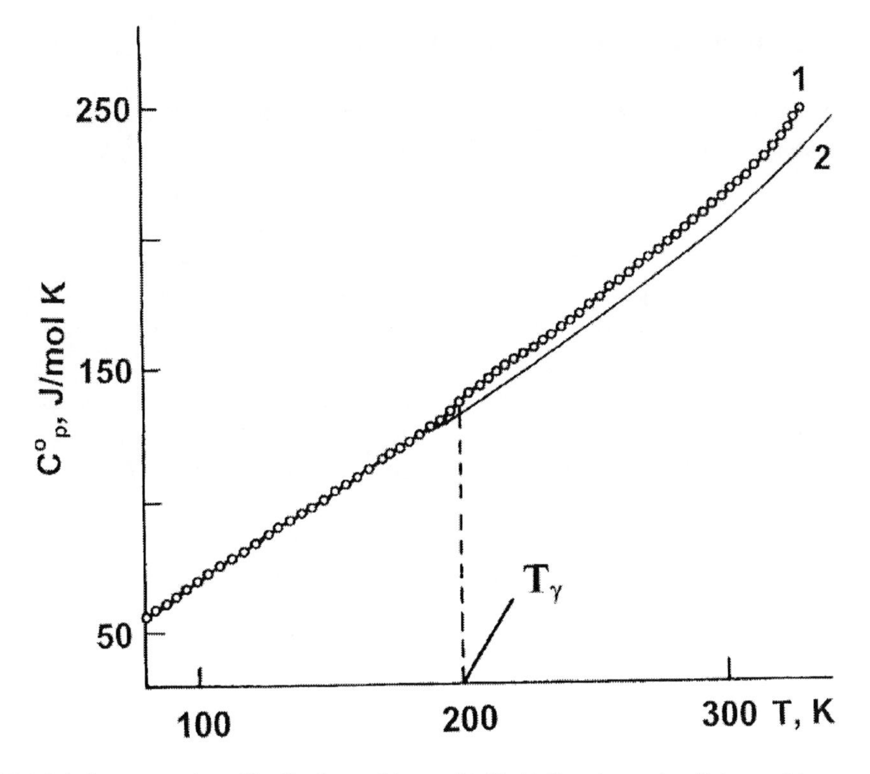

Figure 15. Mole heat capacity of inulin from chicory (1) [4, 134] and wood cellulose with zero crystallinity index (2) [17].

It shows that when the polymer molar mass is higher than 5000 it does not actually affect its C°_{p}. The authors of [134] also note that if the polymer molar mass exceeds 10^4, it does not affect its thermodynamic properties. Standard burning enthalpy Δ_cH^o(solid, 298.15) = −2774.4±4.1 kJ/mol) of inulin from chicory was determined in [4, 133]. Based on the obtained value the standard enthalpy of its formation in a solid state at 298.15 K (Δ_fH^o(solid, 298.15) = −1015.9±4.1 kJ/mol) was calculated.

The results of DTA of dry-air (water content being 1.5 mass%) and dehydrated inulin performed in [4, 135] show that on the thermogram of the source air-dry polysaccharide there appear γ – transition (T_γ=215 K), β–trasnsition (T_β=311 K), devitrification (T_g=326 K) and water evaporation [$T_{vap}(H_2O)$=350 K]. When sorption water is removed from the sample three relaxation periods (T_γ=215 K, T_β=318 K, T_g=341 K) are revealed of the thermogram of dehydrated inulin. It can be seen that water does not affect γ–transition, decreases the temperature of β–transition by several degrees and drastically affects vitrification temperature of inulin. It is attributed to a plasticizing effect of water on inulin. The value of T_γ determined using DTA technique is slightly higher than that determined based on the C°_{p}=*f*(T) curve. It seems to be explained by dynamic conditions of heating in the DTA experiment. Thermal destruction of inulin begins at ~383 K and proceeds in helium atmosphere with energy absorption. It should be noted that contrary to other polysaccharides [3-5, 31, 32, 38, 72-74, 84-114, 116, 121, 123, 125, 130] one devitrification temperature range is revealed for inulin. The temperature of the second devitrification is likely to lie in the range of its thermal destruction.

2.2. Water Solubility Limit in Polysaccharides

Thermodynamic compatibility of polymers with plasticizers, in other words, plastisizer solubility in water, is of both theoretical and practical interest. On the one hand, the higher the concentration of a fluid component in the system (up to the solubility limit), the higher the drop of the polymer vitrification temperature. On the other hand, when a plasticizer is introduced into a polymer in the amount larger than can be dissolved, the plasticizer surplus segregates into a separate phase which can crystallize during cooling or evaporate during heating. It results in deterioration of physical-mechanical characteristics and commodity appearance of the polymer structure. I.B. Rabinovich with his research team [1–5] developed the technique of calorimetric determination of solubility (ω, mass%) of crystallizing low-molecular substances (LMS) in polymers at the temperature of LMS melting or below it.

The idea of the technique is as follows: using calorimetric technique we determine the concentration below which fluid component crystallization in the polymer system is absent. If the amount of the fluid exceeds its solubility, the phase of the fluid component is in equilibrium with the polymer solution since the chemical potential of the fluid in the solution and in the free phase is the same. At $T<T_m$ it is a crystalline phase, at $T>T_m$ the phase is fluid. Determination of melting enthalpy of the mass of corresponding crystals allows calculating the mass of the fluid component phase and, based on the obtained value and the total mixture composition, estimating solubility of the fluid component in the polymer.

Experimentally the aforesaid technique is as follows. Heat capacity of some samples of the analyzed system containing different amount of the fluid component are studied in the temperature range from 4 or 80 K up to T_m of the fluid. Each sample is pre-cooled in the calorimeter up to 4 or 80 K, the cooling rate being $5 \cdot 10^{-2}$ K•c^{-1}; at that the surplus of the fluid component over its solubility either entirely vitrificates or crystallizes. In the process of heating (at the rate of $5 \cdot 10^{-3}$ K•c^{-1}) during heat capacity measurement it devitrificates and the formed supersooled fluid crystallizes at about T_g. The sample is again cooled up to the initial temperature and its heat capacity is measured to determine completeness of the fluid component surplus crystallization. In addition, measurement of heat capacity of the mixture with a crystallized fluid surplus allows revealing possible "reinforcement" of the polymer–fluid vitreous solution with a crystallized fluid component surplus that, in some cases, yields a substantial increase in T_g of the saturated polymer solution till the range of pre-melting of the fluid component surplus phase.

Then, using the technique of continuous energy introduction into the calorimeter from some $T<T_m$ up to some $T>T_m$, the amount of energy in the form of heat (q, J) required for melting the crystals of the free water phase is determined. Based on the ratio of this magnitude to the specific enthalpy of plasticizer melting Δh (J/g) the mass of its crystals [m(cr)=q/Δh] is calculated. Knowing the sample mass (m, g), the mass fraction of the plasticizer in it (C_2,%) and molar masses of the repeated group of the polymer (M_1) and the fluid component (M_2) it is possible to calculate its solubility in the polymer at T_m in mass or molar fractions using Eqs. (2) and (3):

$$\omega, \text{mass\%} = \frac{mC_2 - 100 \bullet q/\Delta h}{m - q/\Delta h} \quad (2)$$

$$\omega\text{, mol\%} = \frac{\omega\text{, mass\%} \bullet M_1}{\omega\text{, mass\%} \bullet M_1 + M_2 \bullet (100 - \omega\text{, mass\%})} \quad (3)$$

In fact, to determine plasticizer solubility in polymers with the aforesaid technique it is sufficient to have one sample provided two– or three-fold melting of the crystallized part of the plasticizer is performed. However, for the reliability of the result to be increased this experiment can be carried out with several individually prepared samples containing different plasticizer surpluses.

The error in determining solubility using this technique is 0.5–2 % (the confidence interval with the probability 95 %).

The mass fraction of the plasticizer in its saturated solution in the polymer (ω) at T_m can be also determined based on DTA data using the technique developed by I.B. Rabinovich and his research team [136]. On the thermograms of the samples of polymer mixtures with the plasticizer at some concentrations of the latter we observe an endothermic peak at T_m which corresponds to the melting of the plasticizer phase which has not dissolved in the polymer. At that, the area of the peak is proportional to the melting enthalpy of the plasticizer amount which is in the form of an individual phase (4) [3-5, 137]. In this case the higher the concentration of the plasticizer which does not dissolve in the polymer, the larger the area of the melting peak on the thermogram.

$$\Delta h \bullet m(LMS) = q = K \bullet S, \quad (4)$$

where Δh is the specific enthalpy of LMS melting (J/g), m(LMS) is the mass of LMS surplus (g), q is the amount of energy in the form of heat spent from melting of the LMS surplus phase (J), K is the coefficient of proportionality which characterizes thermal characteristics of the setup, S is the area of the peak of LMS surplus phase melting (cm^2).

If we divide the area of the endothermic peak of the LMS surplus phase melting by the sample mass we obtain the specific area of the endothermic peak of the free plasticizer phase melting (δ) (Figure 16) [4, 136]. Then the obtained straight line of the δ dependence on the plasticizer concentration is extrapolated from the zero δ and get plasticizer solubility in the polymer (ω). The error of ω determination using this technique is 2–3 %. The advantage of this technique is attributed to the fact that it does not require calibration to determine the proportionality coefficient (K) using Eq. (4). The knowledge of the specific enthalpy of plasticizer melting is not required as well.

Water solubility in some polysaccharides was determined using the calorimetric technique in [2-5, 47-49, 72, 73, 84, 85, 87, 88, 102-106, 116, 121, 123, 125, 130, 133, 135, 138-140] (Table 12). The results obtained using other methods are less reliable. Thus, e.g., water solubility in cellulose [56, 141-143] obtained using the techniques of NMR [141], dielectric losses [142], DSC [143] and sorption [56] varies from 5 up to 30 mass% of H_2O. As an example, Figure 17 shows $C_p^\circ = f(T)$ curves of the potato starch mixture with 58.5 mass% of H_2O (M_{mix}=28.37 g/mol) [4, 73, 125]. This mixture contains a saturated solution of water in polysaccharide and a surplus of a fluid component over its solubility in starch.

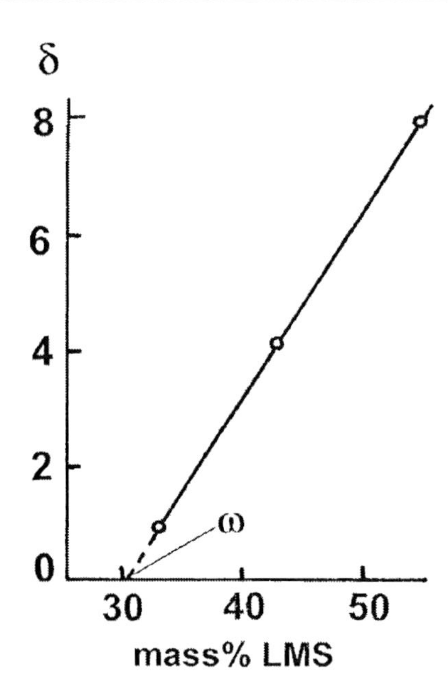

Figure 16. Concentration dependence of specific area of melting endothermic peak (δ) of "free" plasticizer phase [4, 136].

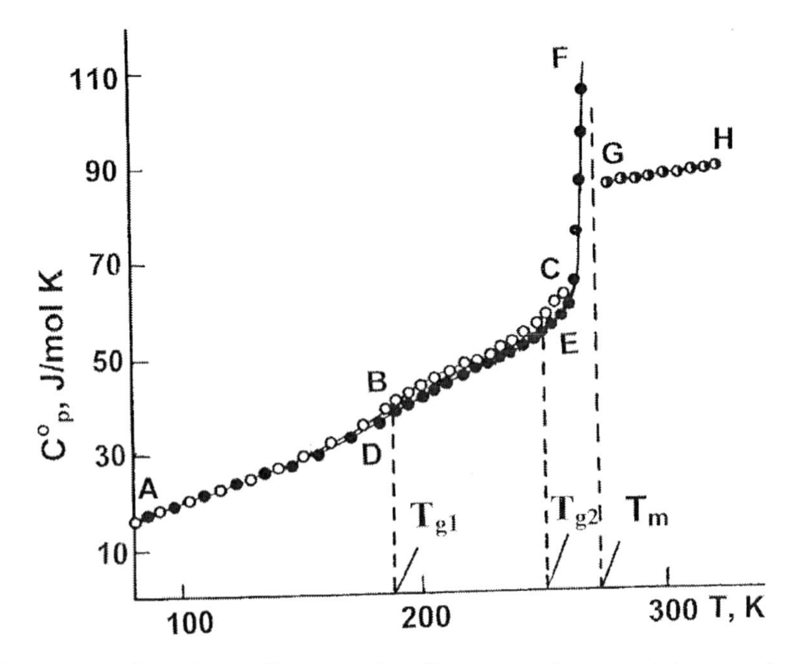

Figure 17. Temperature dependence of heat capacity of potato starch-water mixture sample containing 58.5 mass% of H_2O [4, 73, 125] AB – vitreous saturated solution of water in starch + partially crystalline and vitreous water surplus phase; BC – highly elastic saturated solution of water in starch + partially crystalline and vitreous water surplus phase; AD – vitreous saturated solution of water in starch + entirely crystalline water surplus phase; DE – highly elastic saturated solution of water in starch + crystalline water surplus phase; EF – melting of water surplus phase; GH – highly elastic saturated solution of water in starch liquid water surplus phase.

Table 12. Solubility of water at ~273 K and heavy water at ~277 K in some polysaccharides

Low-molecular substances	Plasticizer solubility in polymer, mass%	References
CELLULOSE (wood, DC*=40%)		
H_2O	26.2±0.2	[139]
D_2O	23.1±0.2	[139]
MICROCRYSTALLINE CELLULOSE (cotton)		
H_2O	15.2±0.1	[47-49]
STARCH (potato)		
H_2O	28.0±0,8	[4, 73, 125]
AGAR		
H_2O	38.0±1.0	[4, 73, 123, 138]
CHITIN (from carposome of hiratake (*Pleurotus ostreatus, Fr.*)		
H_2O	34.0±0.5	[4, 5, 72, 73, 116, 121]
PECTIN (apple, AS-401, DE**=63-65%)		
H_2O	30.9±0.7	[4, 130, 140]
PECTIN (apple, AU-202, DE**=68-76%)		
H_2O	32.4±0.1	[4, 130, 140]
INULIN (from chicory)		
H_2O	17.3±0.7	[133, 135]

*)DC – degree of crystallinity.
**)DE – degree of etherification.

It is apparent from Figure 17 that in the process of initial fast cooling (~20 K/min) of the sample from room temperature up to 80 K the saturated solution of water in starch vitrificated, whereas the phase of the water surplus partially vitrificated and partially crystallized. Heat capacity of the system in the aforementioned state monotonously increases within the range of 80-180 K (section AB). Then the anomalous heat capacity increase (section BC), attributed to devitrification of the saturated solution of water in amorphous (T_{g1}=188 K) and ordered (T_{g1}=250 K) microregions of starch and vitreous fraction of the "free" water phase is observed on the C_p°=*f*(T) curve. A slight exothermal effect due to crystallization of the water surplus from the supercooled fluid state is observed at 260 K.

After crystallization was completed the authors of [4, 73, 125] again cooled the sample up to 80 K and measured its heat capacity. It is apparent from Figure 17 (curve ADE), that the heat capacity values lie below the ABC curve. At that, only saturated water solution devitirifies in the starch microregions with different degrees of ordering. At 273.07 K there is a break in the heat capacity curve which is attributed to melting of the "free" water phase (section EFG). The data of the two experiments on determining water solubility in starch using calorimetric technique are listed in Table 13 [4, 73, 125]. It turned out to be 28±0.8 mass% H_2O (77±0.8 mol.% of H_2O).

Table 13. Data of experiments on determining water solubility (ω, mass%) in starch at 273.07 K for the sample with 58.5 mass% (92.2 mol.%) of water [4, 73, 125]

No. of experiment	Sample mass, g	Enthalpy of "free" water crystal melting, J	Mass of water surplus phase, g	ω, mass%	ω, mol.%
1	3.2802	465.89	1.3967	27.8	76.3
2	3.2802	457.95	1.3720	28.7	77.1
			Average:	28±0.8	77±0.8

As I.B. Rabinovich and his research team noted [2-5, 73], in the samples of polymer mixtures with plasticizers containing an excess of a low-molecular component over its solubility in the polymer the amount of the surplus plasticizer can be so large that its formed crystals will reinforce the saturated solution of the plasticizer in polymer thus preventing it from devitrification. In this case devitrification of the saturated solution will take place in the temperature range of the plasticizer surplus pre-melting phase. The chitin-water sample containing 89.6 mass% of H_2O (Figure 18) [4, 5, 72, 73] can serve as an example.

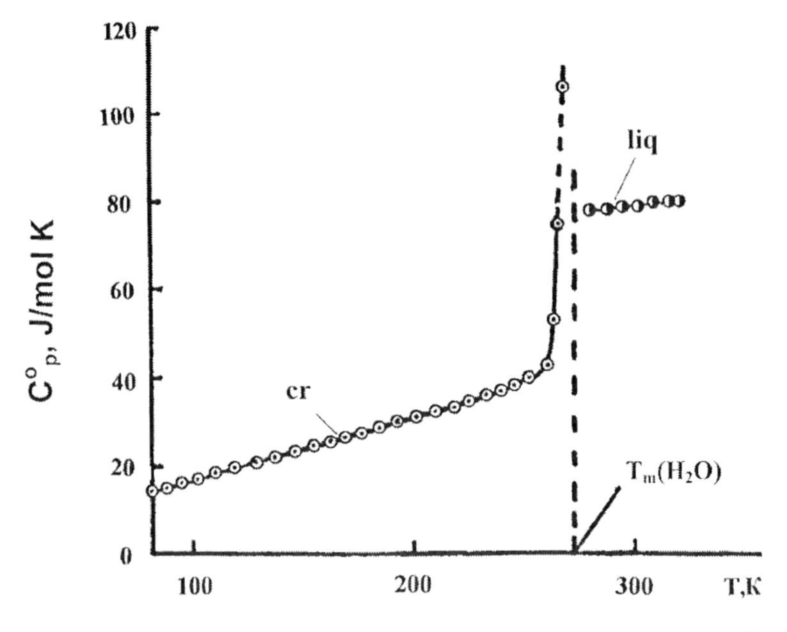

Figure 18. Heat capacity of fungic chitin mixture with 89.6 mass% of water:cr – crystalline water surplus + vitreous saturated solution of water in chitin; liq - liquid water surplus + highly elastic saturated solution of water in chitin [4, 5, 72, 73].

The aforedescribed solubility of H_2O in polysaccharides can be also determined using the DTA technique [3-5, 73, 136]. As an example, we can consider the data on determining water solubility in agar (Figure 19) [4, 73, 123, 138].

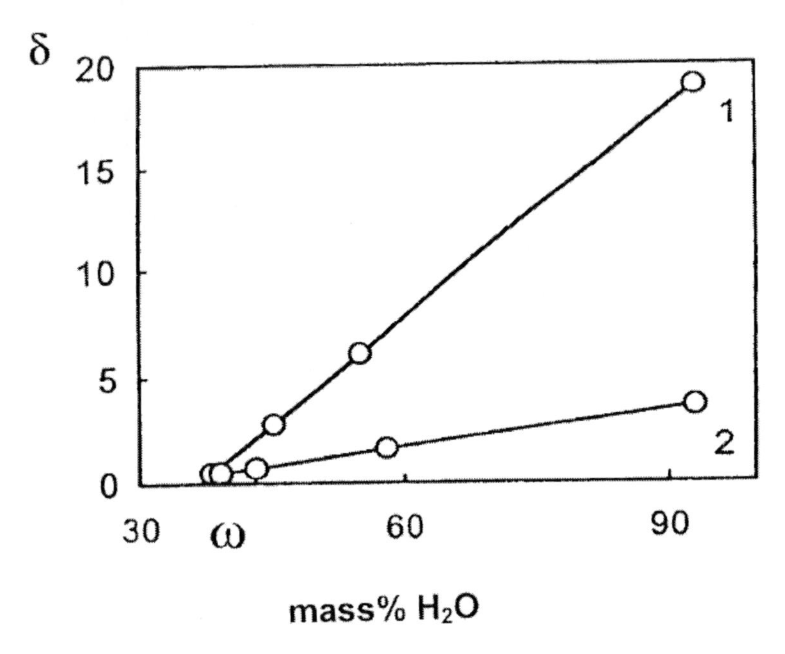

Figure 19. Dependence of specific area of endothermic peak of "free" water melting phase δ on its concentration in mixtures (mass%) obtained for different sensitivity along "ΔT"-axis, μV/cm: 1 – 2; 2 – 10 [4, 73, 123, 138].

The degree of ordering and etherification (Table 12) can affect water solubility in polysaccharides. It is apparent from Table 12 that MCC is more ordered as compared to wood cellulose, and water solubility in AU-202 pectin, characterized by a higher etherification degree, is higher than that in AS-402. Intermolecular interaction in the plasticizer also affects its solubility in a polymer. For example, a comparison of ordinary and heavy water (Table 12) [139] shows that lower solubility of D_2O в D–Cell as compared to solubility of ordinary water in cellulose is mainly attributed to the fact that the energy of O–D•••O bonds opening in heavy water is much higher than that of O–H•••O bond opening in ordinary water [144].

2.3. Diagrams of Physical States of Polysaccharide-Water Systems

Due to the fact that the processes in organisms take place in a water environment it is important for their physiology and biochemistry to study physical states of water in different biologically active substances (BAS), in particular, polysaccharides, as well as the effect of water on temperatures of their physical transitions [8, 16]. The diagrams of physical states of BAS-water systems give valuable information about it [2-5, 73, 145-149]. They allow determining temperature and concentration limits of formation of homogeneous mixtures, i.e. real solutions of H_2O in BAS and BAS in H_2O, as well as two–phase gels, where one of these solutions is distributed in a microdrop way in the matrix of the other. Construction and analysis of such diagrams necessitate, in particular, having the data on temperatures of vitrification of BAS mixtures with H_2O as well as melting temperatures of the phase of H_2O surplus over its solubility in BAS. The data on such type of diagrams can be found in literature. However, the monograph of A.E. Chalykh, et.al. [149] gives the data on the curves of mutual solubility (coexistence curves) in polymer–plasticizer systems and the authors of [145–148] mainly analyze the diagrams constructed based on theoretical statements of polymer physical chemistry. The most complete diagrams of polymer–plasticizer physical states can be found in the papers of I.B. Rabinovich and his research team [1-5, 72, 73, 85, 87, 88, 102-106, 138, 140].

The most characteristic example is the diagram of the starch-water system (Figure 20) [4, 73]. On the diagram: ABC, DSE are the concentration dependences of vitrification temperatures of water solutions in amorphous and ordered microregions of starch, respectively; the sections SE, BC are the vitrification temperatures of saturated water solutions in the starch regions with different degrees of order.

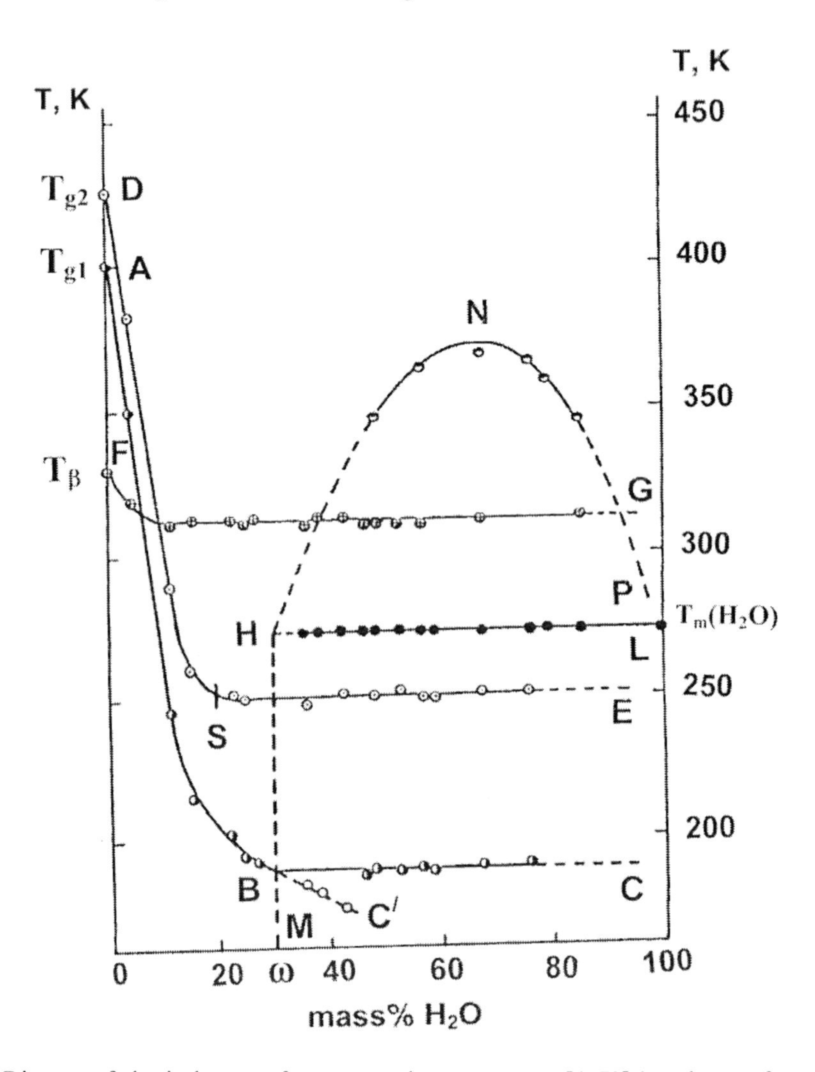

Figure 20. Diagram of physical states of potato starch-water system [4, 73] (see the text for symbols).

It is apparent from the diagram that more ordered (highly associated) regions of starch are saturated with water at a lower total content of water in a biopolymer as compared to less ordered regions (point S is to the left of point B). A similar behavior is also observed in agar-water [4, 73, 138] and pectin-water (Figure 21) [4, 140] systems.

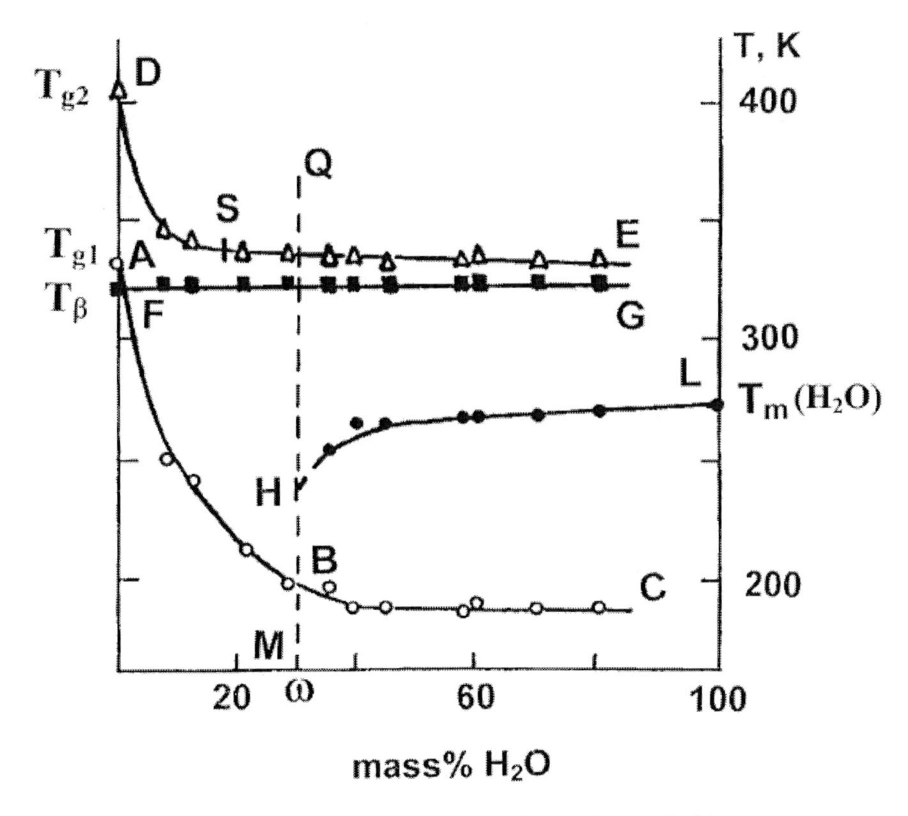

Figure 21. Diagram of pectin-water system [4, 140] (see the text for symbols).

The section of the $BC^{/}$ curve corresponds to vitrification of a nonequilibrium mixture of the vitreous saturated solution of water in starch and the vitreous phase of "free" water. Such system state appears in case of fast cooling (~20 K/min) of the samples containing a minor water surplus. Hereby, the starch-water system is similar to that of chitin-water system (Figure 22) [4, 5, 72, 73] and differs from agar-water [4, 73, 138] and pectin-water [4, 140] systems (Figure 21). FG is the line of temperatures of starch β-transition. This transition is attributed to libration of pyranose rings near the glucoside bond. A slight (15-20 K) decrease in T_β in starch-water and agar-water systems is observed. Starch gels with water are heterogeneous [145–147]. Such gels are formed during decomposition of the polymer solution into two phases. It yields a heterogeneous system whose spatial frame consists of s phase which is rich in polymer, the second equilibrium phase with low starch concentration being positioned in it. A three-dimensional lattice of starch-water gel is formed at the expense of hydrogen bonds. When gel samples were heated from 80 K the crystalline phase of "free" water (line HL on the diagram) melted after devitrification of the saturated solution. In starch-water and agar-water systems $T_m(H_2O)$ remains constant in the region under the coexistence curve (Figure 20, curve HNP) that corresponds to the Gibbs phase rule [150]. In this diagram region three phases (a saturated water solution in a polysaccharide, a highly diluted solution of a polysaccharide in water and "free" water crystals) are in equilibrium at 273 K. Thus equilibrium will be nonvariant. On the contrary, a decrease in $T_m(H_2O)$ is observed in pectin-water and chitin-water systems (Figures 21, 22). It can be attributed, in particular, to the presence of a small amount (fractions of percent) of nonorganic salts, which form eutectic mixtures with water, in these polysaccharides.

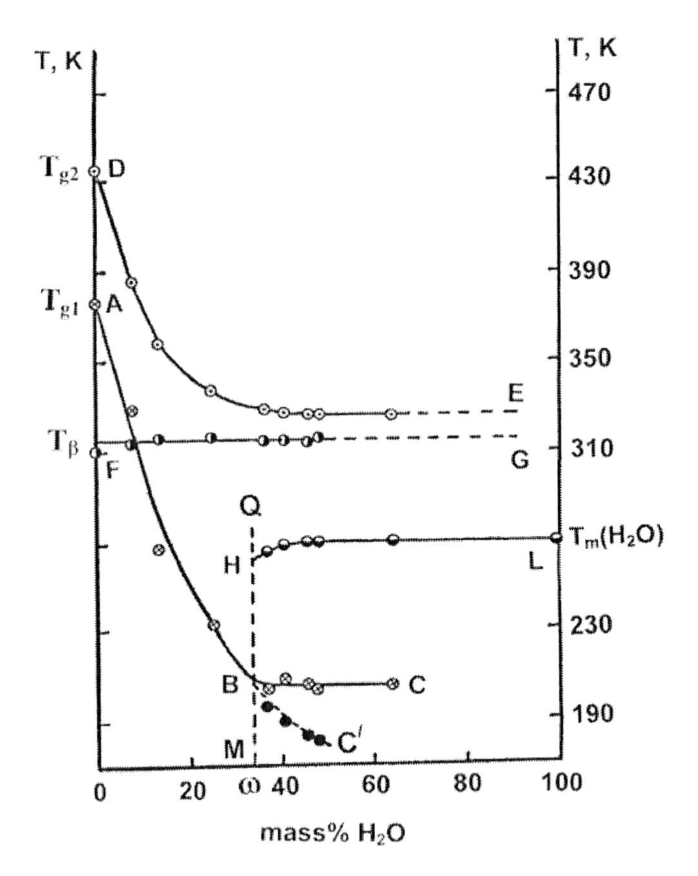

Figure 22. Diagram of physical states of fungic chitin-water system [4, 5, 72, 73] (see the text for symbols).

Decomposition (melting) of the gel appeared in thermograms in the form of a relatively sharp deviation of the base line towards the endothermic region. The agar-water system behaves itself in a similar way (Figure 23) [4, 73, 138]. The melting temperature of the starch gel coincides with the curve of mutual solubility (coexistence curve) (Figure 20, HNP curve). This is the curve with the upper critical temperature of solubilization (UCTS).

The vertical MBH corresponds to the concentration of saturated water solution in starch (28±0.8 mass%). In pectin-water and chitin-water systems it is the MBHQ line (Figures 21 and 22). To the left of this straight line there are homogeneous solutions of water in a polysaccharide; to the right, heterogeneous mixtures of a "free" water phase with its saturated solution in a polysaccharide.

The comparison of diagrams of physical states in starch-water, agar-water, pectin-water and chitin-water systems (Figures 20-22) shows that T_β in the last two systems does not actually depend on H_2O concentration (curves FG, Figures 20 and 21).

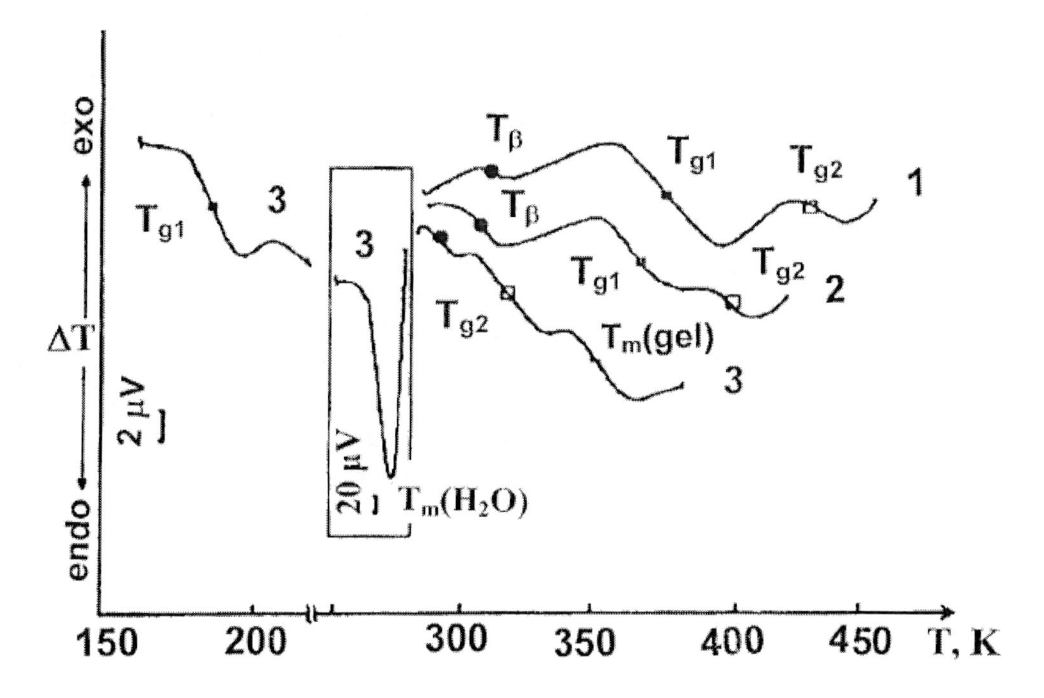

Figure 23. Thermograms of agar samples with different water content, mass%: 1–0; 2–6.7; 3–56.6 [4, 73, 138].

Contrary to the first three systems, more or less ordered chitin microregions in the chitin-H_2O system are saturated at nearly one and the same water concentration (34 mass%) which corresponds to its solubility limit in a polysaccharide.

The analysis of the literary data shows as follows:

- cellulose and other polysaccharides have a complex supramolecular structure. They belong to fibrillary polymers and their microfibrils consist of amorphous and highly-ordered microregions. The result of such structure is as follows: firstly, there are two vitrification temperatures in polysacchcarides. Secondly, several secondary small-scale endothermic relaxation transitions which can be classified as γ- and β transitions are observed in them;
- interaction of polysaccharides (as well as other polymers) with low-molecular substances (plasticizers) is of salvation nature since the process of their mixing is exothermal and the entropic component of this process is negative; under the effect of water a substantial decrease in the vintrification temperatures of both amorphous and ordered polymer microregions is observed;
- A decrease in the polysaccharide vitrification temperatures with an increase of water concentration takes place only until the limit of its solubility in a polymer is reached. Further increase in the water amount yields formation of the second phase in the system, namely a polymer solution in a fluid component which, due to low solubility of polymers in it (hundredths or tens of mass%), represents itself individual water. However, in some cases (e.g., chitin–water and starch–water systems) the temperature of polymer vintrification continues lowering under the effect of LMS and after the limit of its solubility in water is reached (in case the plasticizer did not

have time to decrease the temperature of polymer vitrification to its own temperature and if the cooling is fast) both the saturated solution and the plasticizer surplus vitrify. It seems that in these cases a plasticizing action of the polymer is exerted not only by monolayers of HMB molecules, but also by the second and the third layer of HMB molecules;

- since water can crystallize during cooling, its surplus (*and only it*) is crystallized during cooling and melts during heating. Having determined the amount of energy in the form of heat, required for melting of the aforementioned amount of the fluid component, and knowing its specific melting enthalpy we can calculate the concentration of a saturated water solution in a polysaccharide at its melting temperature;
- the formed system is a two-phase gel where plasticizer microparticles are distributed in the matrix of a saturated solution. In case of large water surplus its formed crystals reinforce the saturated solution in a polysaccharide thus preventing it from devitrification. In this case the saturated solution devitrifies till the temperature of H_2O melting is reached.

Generalizing the literary data on thermodynamic characteristics of some polysaccharides we give the table which lists their thermodynamic functions at 298.15 K (Table 14).

3. Heat Capacity within the Range of 293—323 K of Some Dry Vegetable Products

Nutritive value of plants is often lost in the process of vegetable raw processing. It necessitates developing a technology which would allow for complete preservation of biologically active substances of the source plant. Cryogenic technologies are very promising candidates [151–157]. To optimize the above process technology of manufacturing high-dispersion BAS we should know heat capacity C_p° of the dried vegetable products. The authors of [158] measured heat capacity of fifteen such products in a dry-air state in the range of 293–323 K (Table 15) with the 0.5% error using a modified adiabatic calorimeter designed by S.M. Skuratov [13, 159, 160]. The vegetable products were dried by the authors of [158] at the temperatures from 333 up to 353 K during 6–20 hours. Then they were crushed in an impact crusher. The concentration of residual water in the samples varied from 1.7 up to 8.9 mass% (Table 15). The amount of the residual water was determined based on the mass loss of the sample after its vacuumization (0.6 Pa) at 393 K during one hour. After that the samples mass remained constant.

Heat capacity of air–dry samples of the aforementioned products linearly increases within the analyzed temperature range. It was assumed in [158] that heat capacity of air-dry vegetable products is additive with respect to heat capacities of the dry product and water, the concentration of the latter in the sample being taken into account. The validity of such an assumption is confirmed by the results of [11, 161, 162].

Table 14. Standard thermodynamic functions of polysaccharides
[C_p° и ΔS^o – J/mol·K, ΔH^o и ΔG^o – kJ/mol]. T = 298.15 K, p = 101.325 кПа

Polysaccharide	C_p°	$H^o(T)-H^o(0)$	$S^o(T)-S^o(0)$	$-[G^o(T)-G^o(0)]$	$-\Delta_cH^o$ (s)	$-\Delta_fH^o$ (s)	$-\Delta_fS^o$ (s)	$-\Delta_fG^o$ (s)	$\lg K^o_f$
Agarose $(C_{12}H_{17}O_9)_n$ 306.27 g/mol	354.3±1.1	53.31±0.16	342.3±1.0	48.74±0.15	5735.4±13.3	1559.2±13.3	1758.6±7.0	1034.9±13	181.3
Chitin $(C_8H_{13}0_5N)_n$ 203.19 g/mol	251.9±0.7	37.27±0.11	237.9±0.7	33.67±0.10	4006.4±16.7	999.6±16.7	1253.3±5.0	625.9±13	109.6
Pectin (AS-401) $(C_{6.64}H_{9.28}O_6)_n$ 185.10 g/mol	214.7±0.6	33.24±0.10	214.0±0.6	30.55±0.09	2963.5±3.3	975.7±3.3	1044.8±4.2	664.2±4.9	116.3
Pectin (AU-202) $(C_{6.72}H_{9.44}O_6)_n$ 186.22 g/mol	216.4±0.6	33.66±0.10	218.3±0.6	31.42±0.09	3044.3±4.1	949.3±4.1	1051.4±4.2	635.8±5.3	111.4
Inulin $(C_6H_{10}O_5)_n$ 162.14 g/mol	213.0±0.6	30.51±0.09	190.1±0.6	26.16±0.08	2774.4±4.1	1015.9±4.1	1009.5±4.0	658.2±5.3	115.3
Amylose (AML) $(C_6H_{10}O_5)_n$ 162.14 g/mol	202.5±0.6	29.42±0.09	183.5±0.5	25.29±0.07	2822.9±3.8	967.3±3.8	1013.7±4.0	665.1±5.6	116.5
Amylopectin (AMP) $(C_6H_9O_4)_n$ 145.13 g/mol	180.0±0.5	26.48±0.08	164.9±0.5	22.68±0.07	2522.0±1.7	1125.1±1.7	867.0±3.5	866.8±3.9	151.8
Starch 33.5 мас.% AML +66.5мас.% AMP 150.42 g/mol	187.9±0.5	27.10±0.08	168.7±0.5	23.19±0.07	2616.1±1.6	1075.7±1.6	915.3±3.7	802.4±4.0	140.5

Table 15. Water content in analyzed air-dry vegetable products [158]

No	Product name	Water content,mass%	No	Product name	Water content,mass%	No	Product name	Water content, mass%
I	Black currants *(Ribes nigrum)*	5.4	VI	Bilberry (*Vaccinium myrtillus*)	4.6	XI	Carrot *(Daucus carota)*	7.9
II	Apple *(Malus domestica Bork)*	2.1	VII	Gooseberry *(Grossularia)*	3.1	XII	Cranberry (*Oxycoccus*)	2.8
III	Chokeberry (*Aronia*)	7.9	VIII	Beet *(Beta)*	1.7	XIII	Cucurbit (*Cucurbita*)	3.5
IV	Topinambour (*Helianthus tuberosus*)	5.3	IX	Cherry *(Cerasus)*	4.1	XIV	Cabbage *(Brassica)*	8.9
V	Oats *(Avena)*	7.6	X	Cowberry (*Vaccinium vitisidaea*)	4.6	XV	Parsley *(Petroselinum sativum)*	4.3

Table 16. Averaged heat capacities(J/g·K) of dry vegetable products [158] (for product denomination see Table 15)

T, K	I	II	III	IV	V	VI	VII	VIII	IX	X	XI	XII	XIII	XIV	XV
293	1.475	1.760	1.375	0.710	1.070	1.484	1.410	1.305	1.462	0.900	1.208	1.422	0.920	0.560	0.140
303	1.600	1.880	1.630	0.980	1.170	1.622	1.560	1.492	1.640	1.258	1.350	1.530	1.216	0.820	0.544
313	1.725	2.000	1.880	1.250	1.270	1.758	1.712	1.676	1.830	1.618	1.496	1.640	1.518	1.082	0.945
323	1.850	2.120	2.140	1.520	1.370	1.894	1.860	1.860	2.012	1.976	1.642	1.750	1.815	1.345	1.350

Table 17. Coefficients in approximation equation (5) [158] (for product denomination see Table 15)

Coefficients	I	II	III	IV	V	VI	VII	VIII	IX	X	XI	XII	XIII	XIV	XV
a	1.2250	1.5200	0.8650	0.1700	0.8700	1.2107	1.1100	0.9350	1.0953	0.1827	0.9187	1.2030	0.3233	0.0367	–0.6667
b	0.0125	0.0120	0.0255	0.0270	0.0100	0.0137	0.0150	0.0185	0.0183	0.0358_7	0.0144_7	0.0109	0.0298	0.0262	0.0403

The authors of [158] calculated C_p° of completely dry products (Table 16). It is apparent from Table 16 that the obtained $C_p^\circ = f(T)$ dependences can be approximated with a binomial polynomial

$$C_p^\circ = a + b \cdot t, \quad (5)$$

where t is the temperature in °C.

The coefficients "a" and "b" in Eq.(5) are chosen so that the calculated heat capacities within the range of 292-323 K differed from the experimental ones by less than 0.5% (Table 17).

4. Physical-Chemical Properties of Biologically Active Substances Obtained Using Supercritical Fluid Extraction

Supercritical fluid extraction (SCFE) with carbon dioxide (so called "green chemistry") is a promising technology of vegetable raw processing [163–167]. The process technology of producing extracts with the help of carbon dioxide is ecology-friendly and can substitute extraction of liposoluble substances with organic solvents, namely extraction benzene, hexane, petroleum-ether, etc. Application of liquefied gases for extraction of valuable components from vegetable raw has a number of advantages as compared to conventional methods since CO_2 is nontoxic, incombustible and easy-to-access that is important for commercial production [165, 166]. From the chemical point of view carbon dioxide is inert with respect to the raw components to be extracted [168]. Thus, carbon dioxide can be a base for developing environmentally appropriate manufacturing technologies [169]. Using such techniques the raw (preferably renewable) is processed efficiently, the process is waste-free and no toxic and/or harmful reagents and solvents are used during manufacturing and utilization of the chemical products [163]. According to UNESCO classification the technique of CO_2–extracts manufacturing is nonalternative, ecologically safe, energy- and resource saving, i.e. the one the mankind can take to the XXI century without fearing for its future [163, 164, 168, 169]. Standartization of BAS, produced using SCFE with carbon dioxide, necessitates having reliable data on their physical-chemical properties. A role of standartization as an important factor of scientific and engineering progress was noted by Yu.A. Lebedev and his colleagues [170].

4.1. Supercritical Fluid Extracts from Pine Shoots

Pine is one of the oldest herbs. As far as its phytoncid activity is concerned, it exceeds many wood species. Wood greenery of pine is rich in vitamins both qualitatively and quantitatively [171]. About 70% of needle-free shoots of wood greenery consist of young bark and, as a result, should contain sufficient amount of physiologically active substances characteristic for bioplast. Evidently, wood greenery shoots are BAS sources. The vitamins soluble in organic solvents (and, thus, in carbon dioxide) are the most important compounds of pine wood greenery. A significant role in this vitamin group is played by tocopherol (E vitamin) and K-group vitamins.

Tocopherol concentration in needles increases with tree aging up to a certain limit. Tocopherol concentration is maximum in summer time (17.8%), whereas in winter period it is 4.0% as calculated for the live mass. The bark of young shoots contains a minor amount of this vitamin (from 0.4 up to 0.9 mg% on average). More aged needles contain more E vitamin than the younger ones. Live pine needles contain 2 mg% of K vitamin. The concentration of this vitamin in the bark and shoots is nearly one order lower. The concentration of K vitamin was found to be higher in winter as compared to summer.

The concentration of carotene drastically varies within the annual cycle. As far as needles are concerned, it is maximum during an autumn-winter-spring period (January – 21.31 mg%, May – 19.31 mg%, July – 9.96 mg%). Carotene is mainly found in needles; there is little carotene in bark (up to 1%) and in the wood it is completely absent.

The concentration of A and B chlorophylls in pine needles varies within the annual cycle, i.e. several minima and maxima can be observed. The maximum chlorophylls sum for needles falls at September (184 mg%).

Taking into account that conifer wood greenery is a vegetable raw which can be used fresh the whole year round, it is possible not only to extensively study, but also to use the compounds entering them and possessing miscellaneous properties. These substances find wide application in fragrance-cosmetic and pharmaceutical industries, medicine, etc. [171]. Mixtures of acids from the CO_2–extract of pine wood greenery reveal regulating and fungicidal activity. It allows producing natural pesticides whose ecological harm is minimal. CO_2–extracts of pine wood greenery combine well with the base of creams and shampoos and impart pronounced anti-inflammatory, sedative and wound healing properties to cosmetic products even in low concentrations (up to 0.5%). The extracts smells well and are recommended as a biologically active addition to hair-care products, shaving creams for fat and inflamed skin (e.g. for teenagers) [171]. Pine needle ether oils are used as an aromatizer in soapmaking, fragrance-cosmetic industry, etc. [172].

Using the technique of differential thermal analysis the authors of [173] studied phase and relaxation transitions (melting, crystallization, evaporation, vitrification) in some pine shoot CO_2 extracts with an aim to use the temperatures of the aforementioned transitions as reference points for comparing different lots of the obtained products. In addition, quantitative and qualitative compositions of CO_2 extracts were studied using chromatographic and spectral methods. Pine shoots were extracted by the authors of [173] at T = 313 K and the pressure p=15.0 MPa using the patented technique [174] at the setup developed by the scientific-research company "Biofit" Joint-stock Co. (Nizhny Novgorod). It is noted in [173], that as far as organoleptic indicators are concerned carbon-dioxide extracts from pine shoots represent themselves a thick unctuous mass at room temperature with a characteristic pine smell, the color being from yellow to dark-green. The density of carbon-dioxide extracts is 1.00 – 1.03 g/cm^3. The kinematic viscosity determined with a capillary viscosimeter at 323 K is 900-1400 mm^2/s. Table 18 [173] lists the results of the qualitative chemical analysis of macro- and microelements in pine shoot extracts.

The investigations performed using IR-spectroscopy showed that the spectrum of the hydrocarbon fraction contains 2920, 2960, 2850, 1380, 1460 cm^{-1} absorption bands, typical for C–H oscillations. The IR spectrum of carbonyl compounds reveals 1730, 1250 cm^{-1} absorption bands characteristic for C=O groups. The spectrum of alcohols has 1730, 1250,

1040 cm^{-1}, absorption bands typical for oscillations of C=O and O–H groups of alcohols and ether-alcohols.

Table 18. Results of the chemical analysis of macro- and microelements in pine shoot extract [173]

No	Element	Spectral analysis technique	Result, mg/kg	Technique error, %
1	Calcium	Atomic-absorption	0.00057	10
2	Sodium		0.0025	10
3	Potassium		0.00052	10
4	Aluminum	Atomic-emission	2.60	25
5	Iron		65.0	25
6	Copper		0.25	25
7	Manganese		0.10	25
8	Nickel		0.13	25
9	Magnesium		1.20	25
10	Chromium		0.30	25
11	Tin		0.065	25
12	Molybdenum		<0.40	25
13	Vanadium		<0.10	25
14	Lead		0.10	25
15	Boron		0.18	25

Composition of acids and neutral substances after saponification is shown in Table 19. It is apparent from the table that the extract contains ~30% of neutral substances and ~60% of acids. Using gas-liquid chromatography 9 higher fatty acids and 10 resin acids were identified in carbon-dioxide extracts of pine shoots (Table 20).

Table 19. Results of quantitative separation of carbon dioxide extract of pine shoots [173]

Extracted fraction	Mass fraction, %	
	Before saponification	After saponification
Neutral substances	31.0	24.5
Including		
Hydrocarbons	1.21	1.22
Carbonyl-containing compounds (aldehydes, ketones, esters)	7.00	2.82
Alcohols	14.32	16.12
Polyfunctional compounds	3.41	2.21
Free acids	61.7	65.6
Including		
Higher fatty acids	6.74	10.74
Resin acids	51.86	46.06
Oxidated acids	3.10	8.80

Table 20. Component structure of higher fatty and resin acids entering carbon dioxide extracts of pine shoots [173]

Higher fatty acids	Content, mass%	Resin acids	Content, mass%
11-Methyldodecanoic	0.05-0.08	Pimaric	0.32-2.48
Myristic	0.06-0.38	Ssandaracopimaric	1.20-4.05
Palmitic	0.73-3.15	Palustric and levopimaric	7.29-17.06
Stearic	0.19-0.90	Isopimaric	0.20-0.96
Oleic	0.21-2.07	Abietic	0.70-4.55
Linoleic	0.54-1.81	Dehydroabietic and neoabietic	21.55-37.90
Arachidic	0.17-1.12		
Linolenic	0.10-0.98		
Behenic	0.10-0.94		
Totals:	2.45-8.37	*Totals:*	39.86-67.58

The maximum concentration of F vitamin is 3.22%; that of E vitamin, 10–50 mg/100 g. The content of carotinoids in CO_2-exctracts determined using photometry varied from 18 to 35 mg/100 g; that of chlorophyll, from 140 to 162 mg/100 g of extract. The extracts from pine shoots are characterized by large acid numbers (74-119 mg KOH/g) and relatively small ether numbers (15–60 KOH/g).

The DTA performed in the range of 80–400 K for four samples of CO_2-extracts at different times with an aim to study reproducibility of temperature transitions demonstrated that similar relaxation and phase transitions are found at the thermograms of all samples [173]. As an example, Figure 24 shows the thermograms of one of the analyzed CO_2-extract samples from pine shoots [173]. This sample (as well as other ones) demonstrates an endothermic relaxation transition (T_{g1}) within the temperature range of 240–255 K (Figure 24, curve 1) during initial heating. Then an endothermic peak of water phase melting at 273 K and one more relaxation transition (T_{g2}) within the range of 300–320 K appear followed by two endothermic peaks with two extremums at ~355 and 380 K. At that, a sample weight loss from 6 to 20 mass% is observed. It points to the fact that the last two peaks are attributed to evaporation of water and some volatile organic compounds; at that, the water does not solve in extract components. It is likely that extract components do not also solve in water since its melting temperature does not decrease.

After evaporation of volatile components the samples were cooled and DTA was performed once again (Figure 24, curve 2). The thermogram reveals no peaks corresponding to water melting and evaporation; the signals corresponding to relaxation transitions within the ranges of 240–255 K and 300–320 K are preserved and two more signals (T_{g3} and T_{g4}) with the mean temperatures of 353 and 378 K appear in the region of the evaporation peaks. The recorded relaxation transitions should be attributed to devitrification of the components entering the extracts [117–119].

The amount of water in the analyzed samples [173, 175] was determined based on the area of the water phase melting peak [176]. Both the vegetable raw and the used carbon dioxide can the sources of water in CO_2-extracts. Thus, CO_2 purity is of particular importance.

Thus, devitrification-type relaxation transitions (T_{g1} = 247 K and T_{g2} = 309 K) which are reproducible and whose temperature does not depend of the water present in the samples, are observed in the analyzed samples of pine shoot extracts. The temperatures of these transitions can serve as reference points to standardize the samples of CO_2-extracts from pine shoots.

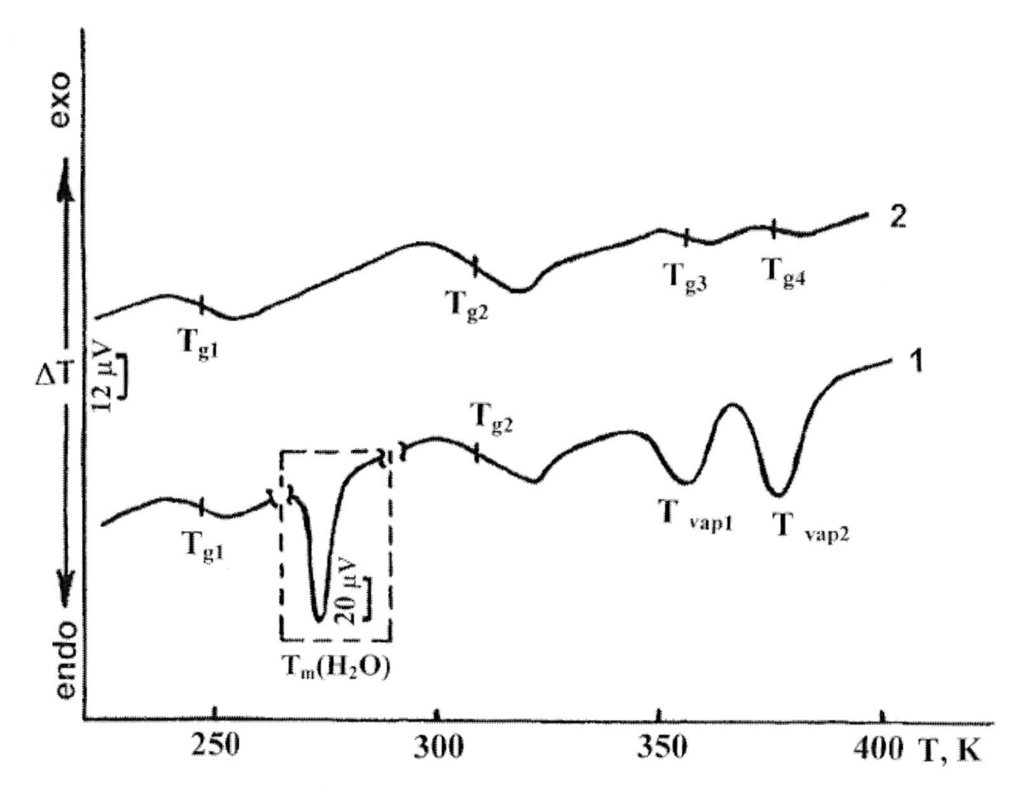

Figure 24. Thermograms of CO_2-pine shoot extract [173] 1- first heating; 2 – reheating after water evaporation.

4.2. Supercritical Fluid Extracts from Weeping Birch (*Betula Pendula*) Bark

A CO_2-extract from weeping birch *(Betula pendula)* bark is a white or a slightly yellowish powder. It contains up to 43 mass% of a valuable substance, namely betulin (synonyms: betulinol, birch camphor, lupendiol). It is a natural BAS representing itself a pentacyclic triterpene alcohol of lupan structure. It exerts antiseptic, antiviral, choleretic, gastro- and hepatoprotrctivr action [177–180]. Its gross formula is $C_{30}H_{50}O_2$; M=442.7 g/mol.

The results of DTA of the weeping birch *(Betula pendula)* bark CO_2-extract (Figure 25) are given in [181]. The experiment aimed at establishing reproducibility of physical and phase transitions in the sample was performed by the authors of [181] three times. The sample was initially cooled with the rate of ~20 K/min from room temperature up to 80 K. Then DTA-curves were plotted in the process of heating with the rate 5 K/min. In case of the first and the second heating (Figure 25, curves 1, 2) the experiment was interrupted at 400 K. Thermograms 1 and 2 demonstrate the appearance of three endothermic relaxation transitions (T_{g1}; T_{g2}; T_{g3}) with the mean temperatures T_{g1}=268 and 269 K; T_{g2}= 319 and 318 K; T_{g3}= 376 and 373 K, respectively, within the range of 260–390 K. In case of the third heating (Figure

25, curve 3) the experiment was ended at 510 K. Thermogram 3 also reveals three physical transitions in the range of 275–390 K (Figure 25, curve 3). However, their temperatures were 3–10 K higher as compared to the first two experiments.

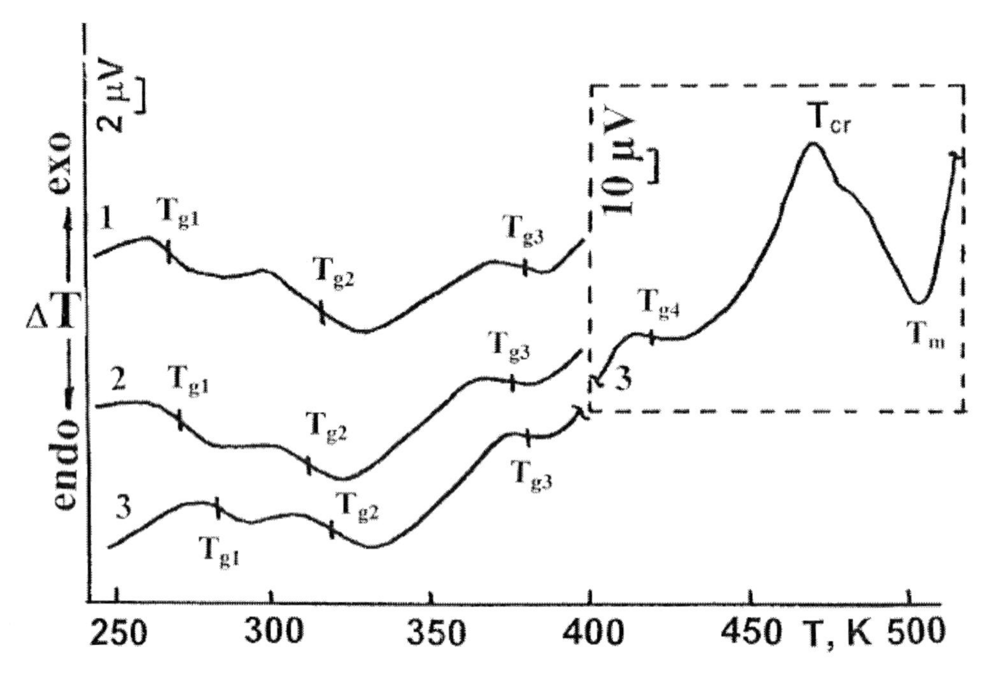

Figure 25. Thermograms of CO_2- weeping birch *(Betula pendula)* bark extract [181] 1- first heating; 2 – second heating, 3 – third heating.

The authors of [181] note that after the first two heating the sample lost ~2% of its initial weight. It could have been adsorption water or other volatile low-molecular components. It is known [1–5], that LMS exert plasticizing action onto polymers thus decreasing the temperatures of their physical transitions. The authors of [181] attribute an increase in T_{g1}, T_{g2} and T_{g3} temperatures to evaporation of LMS from the birch bark CO_2-extract. In addition, the fourth physical transition (T_{g4}) was observed on thermogram 3 in the range of 410–430 K (Figure 25, curve 3). Its amplitude was substantially larger than those of the first three transitions.

After that there appeared a large exothermic peak (T_{cr}=467 K) which can be related to the process of crystallization and an endothermic peak (T_m= 501 K) which is likely to be attributed to melting (Figure 25, curve 3).

The sample weight loss during the third heating made ~0.5 mass%. It points to the absence of destructive processes in the sample [181]. As it was noted above, the extract contains ~43 mass% of betulin. Some of the observed transitions are likely to be attributed to this very BAS [182, 183].

Thus, the authors of [181] recommend to use the temperatures of the first four physical transitions (T_{g1}=272 K, T_{g2}=319.5 K, T_{g3}=378 K and T_{g4}=418 K), which are reproduced well during the sample heating not higher than 430 K, for standartization of the weeping birch *(Betula pendula)* bark CO_2-extract.

5. Bioavailability of Biologically Active Substances Determined with Thermochemical Technique

5.1. Enthalpy of Enzymic Starch Hydrolysis

Enteral tube feeding is an important therapeutic method utilized in case a patient cannot eat himself (herself), i.e. in the result of craniocerebral or maxillary injury. BAS obtained from vegetable raw are used as a component of such diet. Action of BAS mainly depends on the degree of assimilation of useful substances contained in them by a human organism. It necessitates application of an instrumental technique of determining the degree of gastrointestinal digestion in a human organism. Thermodynamic methods [3-8, 122, 184] have been lately used in the solution of biochemical and biological problems. Thus, e.g., attempts have been made to characterize enzymic reactions and processes with participation of different proteins, cells and their assemblages by variations in principle thermodynamic functions, namely entropy, enthalpy and Gibbs functions [185-208]. Experimentally determined enthalpy of enzymic hydrolysis in some polysaccharides and BAS which they enter as well as heat capacity of polysaccharides and enzymes in a wide temperature range allow calculating enthalpy, entropy and Gibbs function of theses substances. Based on the obtained data we can calculate thermodynamic characteristics of polysaccharide enzymic hydrolysis reactions which simulate the processes taking place in a human gastrointestinal tract during digestion. It will give the researchers objective criteria for evaluating the quality of food additions and will allow for deliberate selection of those which efficiently digest and assimilate in the human organism. It should be noted that kinetic regularities of enzymic processes are studied in detail [131, 184, 208], that cannot be said about their thermodynamic characteristics.

The authors of [209, 210] applied the calorimetric technique of determining enthalpy to study the reactions of enzymic hydrolysis of some polysaccharides which model the process taking place in a human gastrointestinal tract during digestion. They determined enthalpy of hydrolysis in the presence of potato starch α-amylase and starch-containing components of "Champion" product intended for enteral tube feeding (developed at "Fitograd" Joint-stock Co., Nizhni Novgorod) [210, 211].

They analyzed potato starch with 13.6 mass% of H_2O in air-dry state. Based on the data of [5, 73, 125] it contained 33.5 mass% of amylose and 66.5 mass% of amylopectin. "Champion" preparation [210, 211] includes the following starch-containing components (mass%): wheat germs (7.76), potato (15.52), apple (23.27) and carrot (3.10). They were prepared using the unique technology of cryogenic grinding developed at "Biofit" Joint-stock Co. (Nizhni Novgorod) [155-157]. The air-dry samples of vegetable raw contained from 4 to 6 mass% of water.

α-Amylasa (α-1,4-glucan-4-glucanohydrolase, KF3.2.1.1), which is also called dextrinogenamylasa or glycogenasa, belongs to enzymes participating in catabolic reactions (decomposition reactions). It is endoamylasa facilitating breaking of intramolecular α-1,4-glucoside bonds of amylose and amylopectin, α-1,6-glucoside bonds of the latter not being affected. Thus in the presence of α-amylasa starch is mainly hydrolyzed up to low-molecular dextrin and some amount of maltose [212-215].

α-AMYLASA

STARCH + H_2O ⟶ MALTOSE + DEXTRIN

One of the models of enzymic reactions is formation of an intermediate short-lived complex between an enzyme and the substance (substrate) which undergoes changes under the effect of enzyme. The reaction being over this enzyme0substrate complex decomposes into product(s) and the enzyme [216].

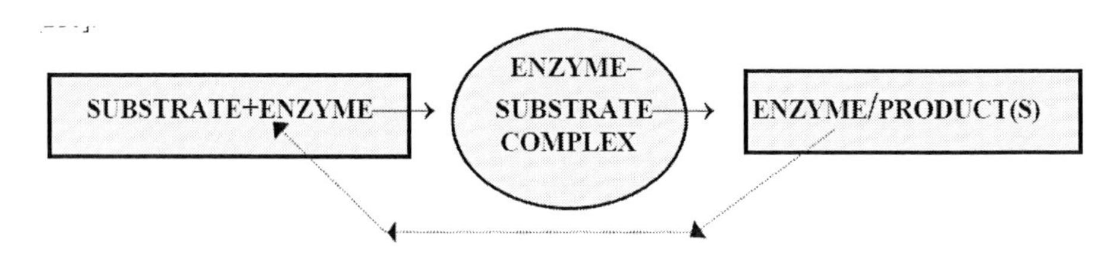

The enzyme was extracted from the cattle pancreas and represents itself a wide powder produced by sublimation drying. Based on the data of [217], alkali medium (e.g., that in the small intestine) whose pH is regulated by sodium bicarbonate contained in human intestinal juice is optimal for amylase. Thus the authors of [210] created pH = 8.3 using $NaHCO_3$. The optimal temperature of enzymic processes lies within the range of 303—313 K [184, 208, 215]. Thus the experiment considered in [210] was performed at 310 K, it being close to the normal temperature of a human body. Amylolytic activity of α-amylase was determined using the calorimetric technique (λ=590 nm), as recommended in [218], with Lachema (AMS-50) biotest. The obtained specific activity of the enzyme was 2260 $E \cdot mg^{-1}$.

Energy effects which accompany hydrolysis of starch and starch-containing vegetable products in the presence of α-amylase was measured in [210] using differential microcalorimeter of Tian-Calvet type [122, 219]. The experimentally obtained differential curves of heat generation and heat absorption carry information on the velocity of energy generation and absorption and the area between the curve and the time axis corresponds to the amount of energy which accompanies the process under study.

Hydrolysis enthalpy (ΔH_{hydr}) [210] for starch and cryopowders in an air-dry state is recalculated for one gram of dry (anhydrous) product. In addition, the authors of [210] compared the obtained data on the hydrolysis enthalpy with the enthalpy of enzymic hydrolysis of the same components which were thermally dried and mechanically crushed. It allowed evaluating the degree of assimilation of the aforementioned products conserved using different techniques. The obtained values were compared with the enthalpy of product mixing with the water solution of $NaHCO_3$ (pH=8.3) without the enzyme (ΔH_{mix}).

Only the data on the enthalpy of starch mixing with water (Table 22) were available in literature [60] and thus the authors of [210] measured the enthalpy of its mixing with a water solution of $NaHCO_3$ (pH=8.3). This process is exothermal. Not only the enthalpy of starch macromolecule hydration with water molecules but also the negative enthalpy of starch transition from a vitreous state (initial) into a highly-elastic one during its mixing with water contribute to the negative value of starch ΔH_{mix} with the water solution of $NaHCO_3$. It is so-called excess enthalpy of a vitreous state [65–69].

Table 21. Determination of water content in pine shoot extracts [173, 175]

Characteristic	*H_2O (distil.)*	*Extract 1*	*Extract 2*	*Extract 3*	*Extract 4*
Sample mass, g	0.1986	0.2149	0.3591	0.2668	0.2274
Peak area, g	0.1671	0.1012	0.0957	0.0231	0.0223
Water mass in sample, g	0.1986	0.1203	0.1149	0.0267	0.0273
H_2O content, mass%	100	56	32	10	12

Table 22. Enthalpy of dry potato starch mixing with water at 300 K [60]

N_2 (starch), mol.%	$-\Delta H_{mix}$, kJ/(mol mix.)	$-\Delta H_2$, kJ/(mol starch)
10	1.76	17.60
20	3.01	15.05
30	3.93	13.10
40	4.60	11.50
50	5.19	10.38
60	5.10	8.50
70	4.52	6.46
80	3.35	4.19
90	2.18	2.42

According to the calculations performed in [60] the aforementioned value is as follows (kJ/mol of polysaccharide): –7.1 for starch, –5.2 for cellulose and –11.7 for dextran. In the experiments performed in [60] the mass fraction of starch in the final solution varied from 2 up to 5 mass%. The value of ΔH_{mix} for the process of starch mixing with the water solution of $NaHCO_3$ within the mentioned concentration range, determined in [210] is –88.0 ÷ –103.5 J/(g of starch). Within the experimental error margin this value coincides with that of the enthalpy of starch mixing with water [60]. The enthalpy of α-amylase (0.6 mass%) mixing with the water solution of $NaHCO_3$ (pH=8.3) is –12.3 J/(g of amylase).

The enthalpy values of hydrolysis of air-dry starch (13.6 mass% of H_2O) in the presence of α-amylase as calculated per one gram of mixture (ΔH_{hydr}) obtained in [210] are shown in Figure 25 (curve 1). The same figure gives the values of ΔH_{hydr}, calculated per one gram of starch (ΔH_{hydr1}) (Figure 25, curve 2) and 1 gram of amylase (ΔH_{hydr2}) (Figure 25, curve 3).

The authors of [210] introduced a correction for water contained in air-dry starch (13.6 mass%) and recalculated the obtained values of ΔH_{hydr} for anhydrous starch. The obtained results are listed in Table 23. It is apparent from Figure 25 and Table 23 that the process is exothermal and the value of ΔH_{hydr} is ~3 times smaller that the enthalpy of starch mixing with water [60]. The resultant value of ΔH_{hydr} includes negative enthalpy of the process of starch mixing with water, positive enthalpy of glucoside bond breaking in macromolecules of amylose and amylopectin, enthalpy of the process of terminal hydroxyl group formation in the products and negative enthalpy of hydration of the formed products.

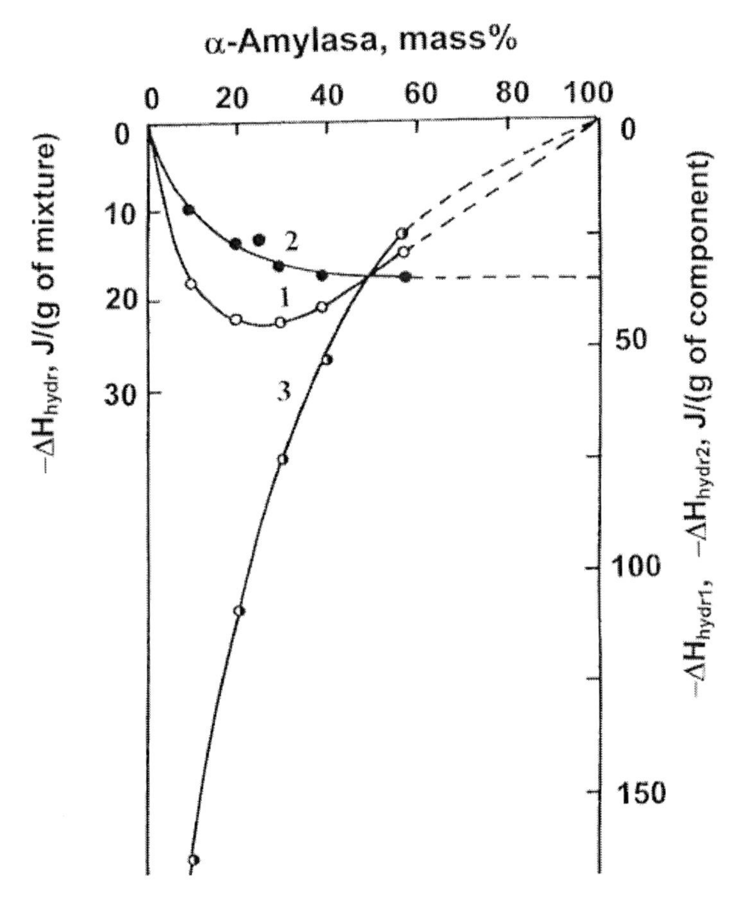

Figure 26. Enthalpy of enzymic hydrolysis of air-dry starch, calculated per g: 1 – mixture (ΔH_{hydr}), 2 – starch (ΔH_{hydr1}), 3 – amylase (ΔH_{hydr2}) [210].

5.2. Enthalpy of Enzymic Hydrolysis of Components of Vegetable Raw Products for Enteral Feeding

Tables 24 and 25 [210] list the results of determining the enthalpy of hydrolysis with the water solution of $NaHCO_3$ (pH=8.3) in the presence of α-amylase of vegetable products and mixing of some of them with the water solution of $NaHCO_3$ without α-amylase. It is apparent from Tables 24 and 25 that these values are negative. At that the absolute value of ΔH_{hydr} is larger than ΔH_{mix}. This difference exceeds the errors of enthalpy determination. The fact that in this case an inverse tendency is observed (as compared to starch) points to more complex processes which accompany enzymic hydrolysis of starch in natural vegetable products. Based on the data of [220] α-amylase attacks a starch grain in a vegetable product, loosens its surface thus forming channels and grooves, i.e. splits the grain into parts. We think that it allows evaluating a degree of starch accessibility in vegetable raw depending on its processing based on hydrolysis enthalpy. It is apparent from Tables 24 and 25 that the values of ΔH_{hydr} and ΔH_{mix} for the sample made of ground wheat germs are close, i.e. starch in the ground sample is practically inaccessible for the enzyme.

Table 23. Experimental enthalpy values of dehydrated starch hydrolysis with water solution of $NaHCO_3$ (pH=8.3) in the presence of α-amylase at 310 K [210]

Starch mass, g	Amylase mass, g	Mass fraction of amylase, mass%	Q_{total}*, J	$-\Delta H_{hydr}$, J/(g mix.)	$-\Delta H_{hydr1}$, J/(g starch)	$-\Delta H_{hydr2}$, J/(g amylase)
0.1662	0.0233	12.29	3.8206	20.16	22.99	164.0
0.1736	0.0501	22.40	5.4984	24.57	31.67	109.7
0.1443	0.0715	33.13	5.3896	24.97	37.35	75.38
0.0679	0.0503	42.55	2.7024	22.86	39.79	53.72
0.0472	0.0548	53.72	1.8863	18.51	39.96	34.42
0.0790	0.1238	61.07	3.1109	15.34	39.39	25.12

*) Q_{total} – Amount of energy released during experiment.

Table 24. Hydrolysis enthalpy (ΔH_{hydr}) of cryopowders of crushed vegetable products and "Champion" preparation in the presence of α-amylase at 310 K [210]

Product	$-\Delta H_{hydr}$, J/(g of dry product)	Product	$-\Delta H_{hydr}$, J/(g of dry product)
Wheat germs (cryo.)	41.8	Apple (crushed)	20.9
Wheat germs (crushed)	36.8	Carrot (cryo.)	14.2
Potato (cryo.)	38.9	Carrot (crushed)	10.5
Potato (crushed)	25.9	«Champion»	13.8
Apple (cryo.)	23.0		

It is important that the enthalpy of cryopowder hydrolysis is more negative (by 1.2–1.5 times) as compared to thermally dried and mechanically crushed samples. It shows that starch is cryopowders is more accessible for the enzyme as compared to mechanically crushed vegetable products. Thus it can be assumed that both cryopowders and "Champion" preparation containing them will better digest and assimilate in the human organism. The authors of [210] calculated ΔH_{hydr} of "Champion" preparation based on additive contributions of hydrolysis enthalpies of the cryopowders contained in it taking into account their concentration in the preparation. It turned out to be −14.6 J/(g of dry product). This value is close to the experimentally obtained one, namely $\Delta H_{hydr} = -13.8$ J/(g of dry product) (Table 25). Thus the experiments performed in [210] show that a degree of assimilation of products intended for enteral feeding can be estimated based on additive contributions of enzymic hydrolysis enthalpies of the components contained in them.

Table 25. Enthalpy of mixing (ΔH_{mix}) cryopowders with crushed vegetable products as well as that of "Champion" preparation with water solution of $NaHCO_3$ (pH=8.3) at 310 K [210]

Product	Wheat germs (cryo.)	Wheat germs (crushed)	Potato (cryo.)	Potato (crushed)	Apple (cryo.)	Apple (crushed)	Carrot (cryo.)	Carrot (crushed)	«Champion»
$-\Delta H_{mix}$, J/(g of air-dry product)	37.2	35.1	33.9	22.2	19.1	17.6	10.6	7.83	9.20
$-\Delta H_{mix}$, J/(g of dry product)	39.7	37.2	36.0	23.4	19.9	18.4	11.3	8.20	9.62

6. Practical Application of Physical-Chemical Data

6.1. Thermochemical Investigation of La-Bacilli Cultivation on Different Nutrient Media

A complex system of standartization for medical biological preparations, bacterial nutrient media included, presupposes application of unified methods of evaluating the quality of ready preparations [221, 222]. The authors of [221, 222] propose to evaluate the quality of nutrient media based on the microorganism growth curve (Figure 27 in [223]). The plot of microorganism number dependence on time is S-shaped. The growth curve can be divided into four sections (Figure 27): the initial stage (a lag-phase) is characterized by a minor growth; the second stage (a logarithmic phase) characterized by intense exponential growth (the growth rate reaching its maximum); after that the growth rate starts decreasing and the third stage, namely self-deceleration phase begins; the fourth final stage is stationary (a plateau) when the overall growth stops and the population number remains constant.

Growth curves in biological practice are plotted using a colorimetric technique which lies in determining optical densities of dispersion with bacteria with a photoelectrocolorimeter. However using this technique it is difficult to distinguish live and dead bacteria, it is very time-consuming and is not free from subjective errors.

Based on the above said, the authors of [4, 224] made an attempt to utilize a thermochemical technique to study the process of La-bacilli (LB) cultivation on different nutrient media with an aim to develop an objective instrumental method of evaluating their quality. The above technique can be applied to solve the problem stated since the process of microorganism reproduction is accompanied with energy generation [225] in the form of heat. In literature there are publications concerning calorimetry application in biology [6-8, 122, 184-210, 226, 227].

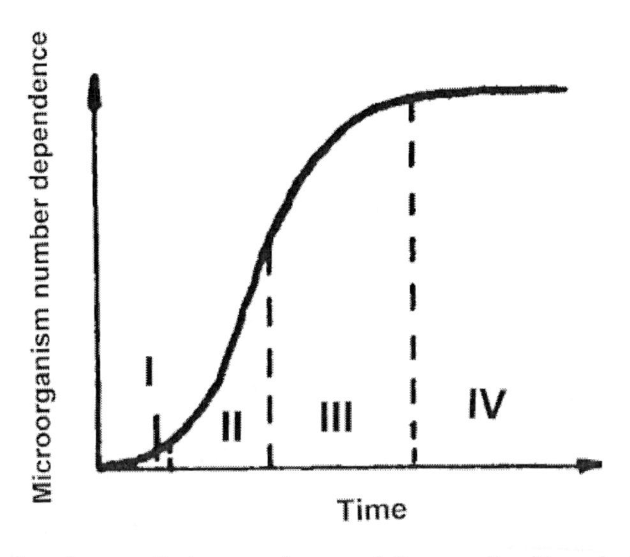

Figure 27. Typical S-shaped curve of microorganism population growth with marked growth phases: I – lag-phase; II – logarithmic phase; III – deceleration phase; IV – stationary phase (plateau) [223].

LBs are intestinal parasites belonging to facultative anaerobes or microphiles, i.e. they do not die in oxygen environment but can do without oxygen. The following nutrient media were chosen in [4, 224] for LB cultivation: casein-yeast medium (CY-5), milk-malt medium (MM), milk-yeast medium (MY) as well as the medium of Moser-Rogos-Sharp without glucose (MRS) and with glucose (MRS+Gl). The amount of energy (enthalpy, $\Delta_{cult}H$) generated in the process of LB reproduction on different nutrient media was measured by the authors of [4, 224] with the help of differential microcalorimeter [122, 219]. LBs in a saline (0.9 mass% of NaCl in water) were sealed in thin-walled glass ampulas whereas nutrient media were poured into pre-sterilized metal ampulas with Teflon plugs. When thermal equilibrium between calorimeter cells was reached (at 310 K) the reagents were mixed by breaking the glass ampoule and extensive mixing. The authors of [4, 224] introduced corrections into the obtained energy value for the energy effects which accompany ampoule breaking, reagent mixing and solvent evaporation.

The results obtained in [4, 224] are shown in Figures 28 and 29. It is apparent from the figures that the process of LB cultivation on the analyzed nutrient media is accompanied with energy generation. The shape of the curves of energy generation [Q, J/(g of product)] time dependence (Figure 28) is similar to the curves of microorganism population growth (Figure 27) and the same sections as those of the growth curves can be distinguished. Thus, an experimenter gets a high-accuracy objective instrumental (thermochemical) technique of constructing growth curves.

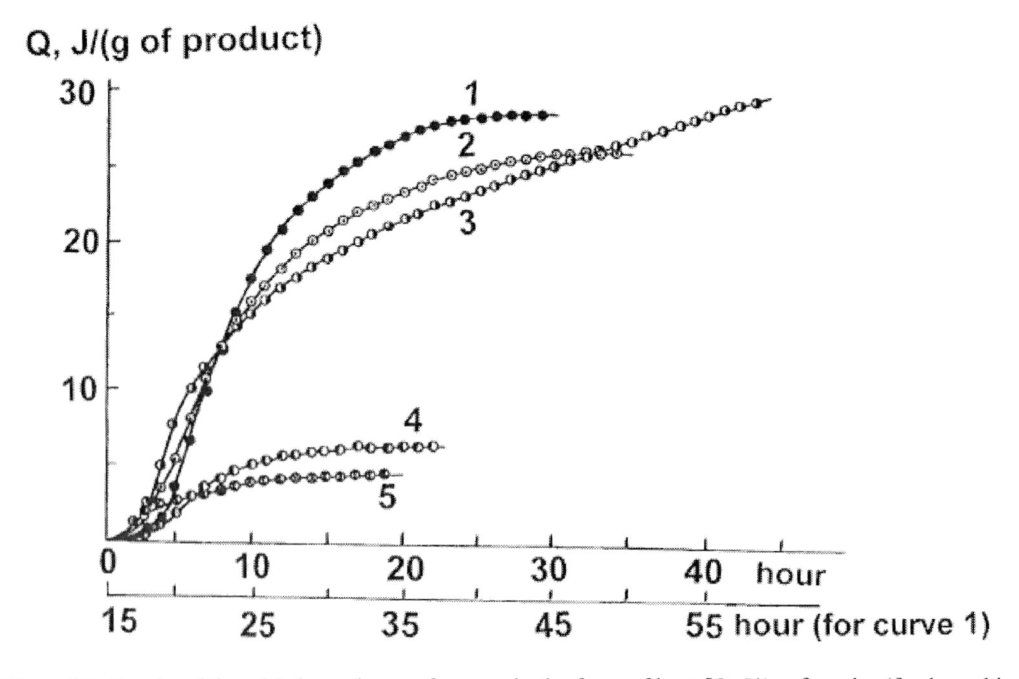

Figure 28. Graphs of time (τ) dependence of energy in the form of heat [Q, J/(g of product)] released in the process of LB cultivation on different nutrient media: 1 – MRS + Gl; 2 – MRS; 3 – MM; 4 – MY; 5 – CY [4, 224].

According to Gordon K. Kresheck [6], the exothermal nature of microorganism cultivation process is attributed to the fact that the initial overall reaction in cells can be expressed by Eq. (6):

$$C_6H_{12}O_6 + 6\,O_2 \rightarrow 6\,CO_2 + 6\,H_2O + \Delta_rH \quad (6)$$

The experiments performed in [4, 224] demonstrated that the amount of the generated energy ($\Delta_{cult}H$) depends on the nutrient medium type. It is maximal in case of MRS+Gl and minimal for CY. Removal of glucose from the MRS nutrient medium yields a decrease in the amount of the generated energy. Since LBs were replanted onto similar nutrient media it is not necessary for them to adapt to new conditions and thus the lag-phase on the obtained curves was absent. In one case when the break between LB growing and the experiment performed in a calorimeter was longer the bacilli became disaccustomed to the medium and a lag-phase appeared on the heat generation curve. Thus only live LB can be registered with a calorimeter. The time dependence of the thermal power which accompanies the process of LB cultivation is extremal (Figure 29), the time of reaching maximal power being different for different nutrient media. According to the recommendations of [221, 222], the large the biomass increase during a shorter period of time, the better the nutrient medium quality. These indicators correspond to the height of the extremum and the time of its reaching (τ_{max}) on the $W=f(\tau)$ curve. The lower the τ_{max}, the better the medium adaptability to trophic (nutrient) demands of LB. The higher the extreme value of the thermal power, the large the substrate consumption and, as a result, the greater the biomass increase.

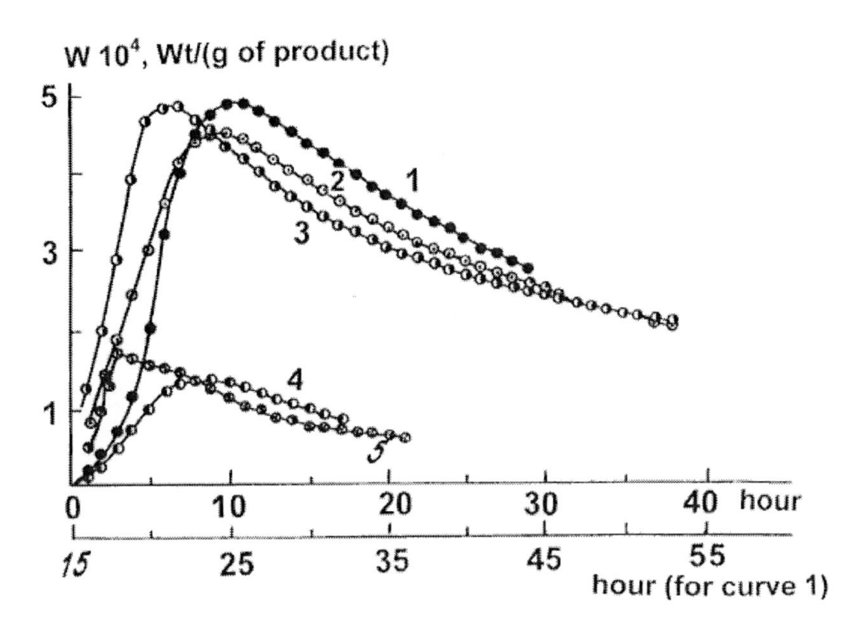

Figure 29. Graphs of time (τ) dependence of power[W, Wt/(g of product)] of the process of LB cultivation on different nutrient media: 1 – MRS + Gl; 2 – MRS; 3 – MM; 4 – MY; 5 – CY [4, 224].

6.2. Methods of Diagnosing Malignant Neoplasms

As it was shown above, both calorimetry and DTA are very informative methods when investigating physical-chemical properties of BAS. They can be successfully applied to solve medical problems, in particular, to diagnose malignant neoplasms [228-233]. Thus, the results of the investigation of physical-chemical properties of human blood erythrocytes and their components (membranes, cholesterol) are given in [234]. For the erythrocytes to be obtained the authors of [234] centrifugated fresh blood at 3000 g during 10 minutes and the erythrocyte sediments were washed four times with 0.9% water solution of NaCl [235]. To separate the membranes, 150 ml of 10mM water solution of NaCl, containing 1 ml of ethylenediaminetetraacetic acid, was added to 25 ml of washed erythrocyte sediments and mixed during 5 min at 273 K. Then the membranes were washed four times with 0.9%-water solution of NaCl and centrifugated at 12000 G and 273 K [235]. The authors of [234] charged the erythrocytes and the membranes into open aluminum crucibles and cholesterol was placed into sealable glass crucibles. The obtained results are shown in Figures 30, 31.

The thermograms of initial air-dry samples of erythrocytes (Figure 30, curve 1) and membranes (Figure 30, curve 2) revealed endothermic peaks of residual water evaporation [$T_{vap}(H_2O)$] at 399.5 and 391 K, respectively. After its direct evacuation from the setup chamber the authors of [234] determined based on the sample mass loss using DTA that air-dry erythrocytes contained 10 mass% of water; the membranes, 7 mass% of water. The thermograms of dried erythrocytes (Figure 30, curve 4) demonstrate endothermic relaxation transitions (T_1, T_2, T_3). They are attributed to an increase in the heat capacity of substances in the result of excitation of separate macromolecule sections entering the erythrocytes. Such transitions were also observed in polysaccharides [3-5, 31, 32, 38, 72-74, 84-114, 116, 121, 123, 125, 130, 133, 135, 138-140], as well as in globular and fibrillar proteins [161, 162, 236-238]. The endothermic phase transitions (T_{tr}) is also observed on the thermogram of dry erythrocytes at 349.5 K.

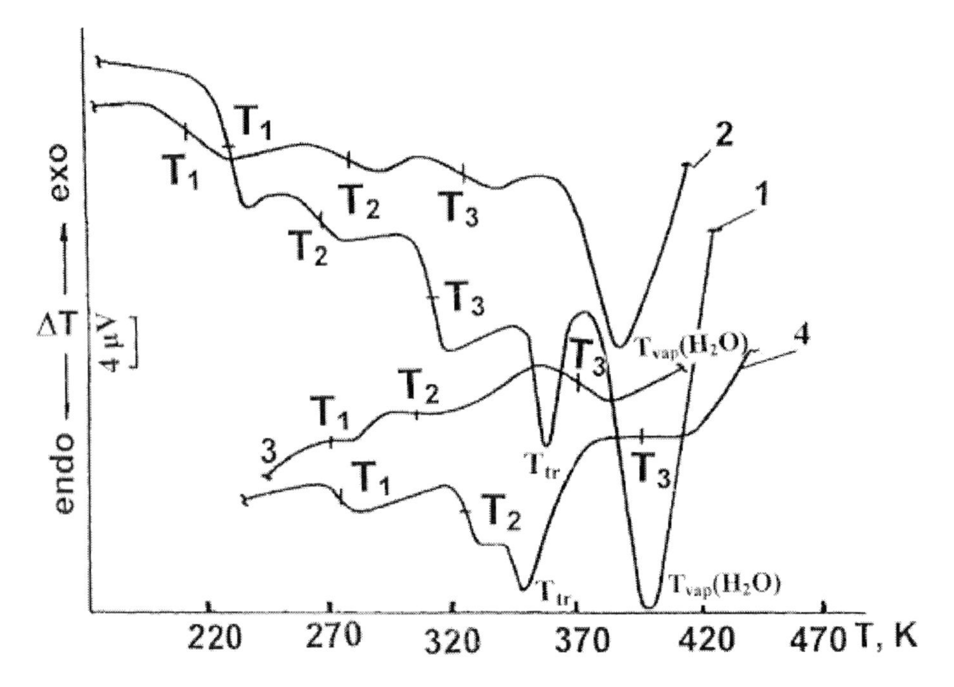

Figure 30. Thermograms of erythrocyte (1, 4) and membrane (2, 3) samples: dried (3, 4) and containing water (1, 2), mass%: 1 – 10, 2 – 7 [236].

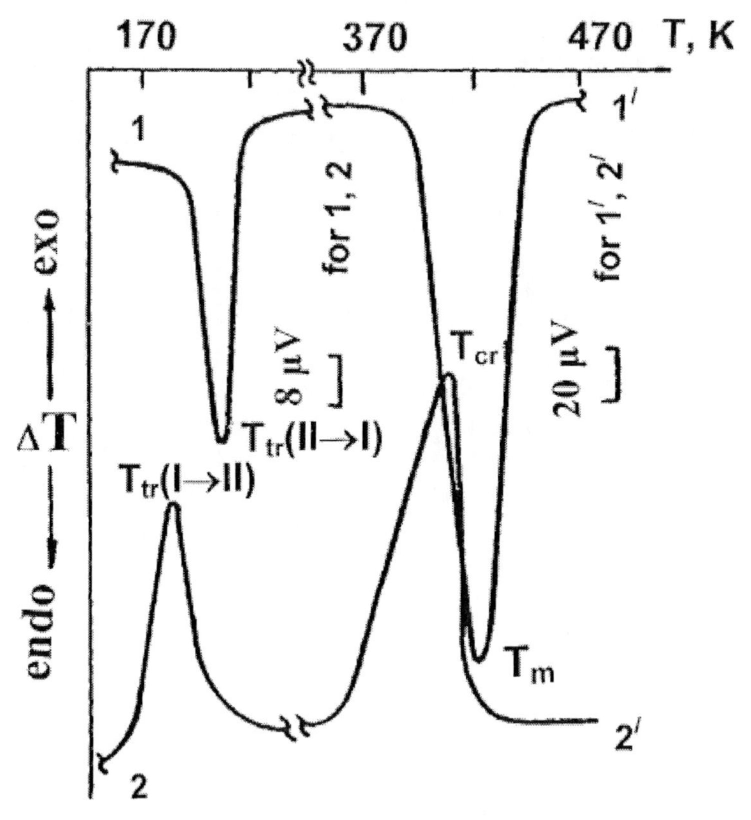

Figure 31. Thermograms of cholesterol obtained during sample heating (1, 1′) and cooling (2, 2′) [236].

The air-dry erythrocytes containing 10 mass% of water are concerned (Figure 30, curve 1) demonstrate the same transitions as the dry ones but with lower temperature and increased amplitude. Plasticized polymers behave themselves in a similar manner [1–5, 145–149]. Thus it can be assumed that with respect to high-molecular erythrocyte components water acts as a plasticizer. The area of the endothermic phase transition for air-dry erythrocytes increased as well.

Membranes are one of the erythrocyte components. The thermograms of dry membranes (Figure 30, curve 3) also revealed several relaxation transitions (T_1, T_2, T_3) that confirms the membrane ability for structural recombination not only when they are in intact cells, but also in an isolated state. Water also exerts plasticizing action on high-molecular membrane components, it yielding a temperature decrease and amplitude increase of these transitions (Figure 30, curve 2). The comparison of the thermograms of dried erythrocytes and membranes (Figure 30, curves 3 and 4) allows making a conclusion that relaxation transitions in them are analogous [234].

Cholesterol, which enters membranes as a component and forms a complex with proteins, belongs to the substances which are capable of getting over to a liquid-crystal state [239, 240]. On its thermogram obtained in [234] during heating (Figure 31, curve 1, 1′) two endothermic phase transitions, namely [T_{tr}(KII→KI) and T_m] are observed. Both of them are reversible since on the thermogram obtained during cooling two exothermal peaks [T_{cr} and T_{tr}(KI→KII)] are observed (Figure 31, curve 2, 2′).

The transition with a greater value is related to cholesterol melting (T_m=422.3 K). This temperature is in agreement with the literary data [241]. Another peak can be attributed to the "crystal II ⇔ crystal I" transition [T_{tr}(KII⇔KI)=306 K]. The authors of [234] also determined the enthalpies of the aforementioned phase transitions [ΔH_m and ΔH_{trans}(KII⇔KI)] using the automated differential calorimeter working on the principle of a triple thermal bridge [242, 243]. A relative error in determining enthalpy of phase transition using the above technique is ±10%. ΔH_m=25.2±2.5 J/g, ΔH_{tr}(KII⇔KI)=2.5±0.25 J/g.

Thus, the demonstrated investigations results showed that the DTA technique can be successfully used to study human blood components. Thus the authors of [244] used it to develop a novel express-method of diagnosing malignant neoplasms. The task of the technique proposed in [244] is as follows: to have a possibility of diagnosing malignant neoplasm in any human organ without traumatizing the organ tissues, to increase its safety, to decrease duration of the analysis and to expand the arsenal of methods intended for malignant neoplasm diagnostics. The problem stated in [244] was solved as follows. Blood serum served as a biomaterial. After it was dried in the air the residual water was removed in vacuum at 0.6 kPa and 383–403 K. Then the dehydrated sample cooled up to room temperature was studied with DTA. To do this it was placed in the DTA setup chamber, first cooled in the inert gas atmosphere first up to 80 K and them heated up to 433–443 K with simultaneous recording of the sample relaxation transitions within the temperature range of 113–443 K. In case relaxation transitions with the mean temperatures <273 K were revealed, a malignant neoplasm was diagnosed [244].

The fact that blood serum was taken as a biomaterial in [244] allowed getting the information on pathology, in particular, on a malignant neoplasm, since blood passes through all organs and tissues and thus actually contains all the substances extracted in case a malignant tumor is formed and develops [228-233, 245, 246]. Contrary to organ tissue sampling blood sampling from a vein does not require operative intervention and is a less invasive and a less traumatic method. Removal of residual water from the blood serum sample, pre-dried in the air, allows avoiding its effect on the temperature of relaxation transitions and in case the serum sample is heated from 113–443 K it allows getting additional information on the development of malignant neoplasms in different human organs based on a relatively sharp increase in heat capacity within a specified temperature range (relation transition). The analysis lasts 30–60 min. The experiment was carried out in the inert gas atmosphere. The temperature measurement error was 0.5 K [244].

The results of the experimental investigation of blood serum of a large number of patients (Figure 32) allowed making a conclusion that relaxation transitions with the mean temperatures of <272 K in the sample of dehydrated (free from residual water) blood serum, which were preliminary cooled up to 80 K and then heated up to 433–443 K, are typical for a patient with a malignant neoplasm [244].

Using the proposed technique the authors of [244] diagnosed 35 patients at Nizhni Novgorod City Oncologic Dispensary (NCOD). In addition, blood serum of 30 physically healthy donors from Nizhni Novgorod Regional Blood Center (NRBC) was analyzed. As the result, the relaxation period at <273 K was revealed on the DTA-curve in 35 patients of NCOD (Figure 32, curves 1m 2) and malignant neoplasms were diagnosed. As far as physically health people were concerned, such transition was absent (Figure 30, curve 3). In addition, concurrently with utilization of the proposed technique neoplasms of all 35 patients of NCOD were diagnosed using standard clinical methods, which also revealed their malignant nature. Thus, the accuracy of the proposed technique made 100%.

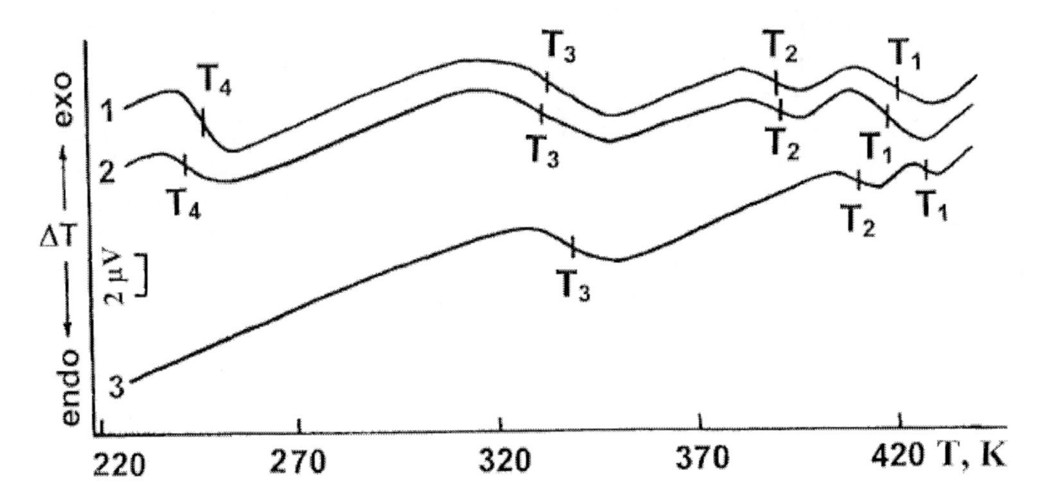

Figure 32. Thermograms of dried blood serum samples of patients A and B (carcinoma of lung) (1, 2) and healthy patient K (3) [246].

It is apparent from the obtained data that the technique proposed in [244] allows diagnosing malignant neoplasms in a human organism with a high degree of accuracy. It is and an express low-traumatic and low-invasive method. In addition, it expands the arsenal of the methods intended for diagnosing malignant neoplasms.

7. Sorption Properties of Polysaccharides

Nowadays pollution of the environment with heavy metals is considered one of the key problems of ecology and health of our planet population. An injurious effect of nonorganic compounds of heavy metals on human health and their high stability in environmental objects initiated a search for treatment-and prophylactics methods intended for decreasing accumulation of heavy metals in the organism and weakening of the pathological changes caused by them. The idea of bioprophylactics is as follows: organism reactivity and resistance of the organism to hazardous factors of the environment (i.e. biological prerequisites for the development of professional or ecology-induced pathology) are fought with rather than the hazardous factors themselves [247, 248]. Enterosorption is an integral part of bioprophylactics. Enterosorbents made of vegetable raw do not only lead heavy metals out of the human organism, but also exert prophylactic action by supplying vitamins, dietary fibers and other health-giving substances. Vegetable enterosorbents whose sorption properties have been intensively studies lately are of particular interest [249-252]. A search for effective enterosorbents capable of lowering the concentration of heavy metals in a human organism up to an allowable level and "softly" affecting a man's organism are of current importance. An ideal enterosorbent should be nontoxic, atraumatic for mucous membranes, easy-to-evacuate from the bowels and possess high sorption capacity with respect to the extracted chime components. It is desirable for the application of nonselective enterosorbents to yield a minimal loss of useful ingredients. The most efficient way of introducing the aforementioned substances is peroral when sorption begins in the stomach and ends in the small intestines. When passing along the bowels the bound components should not be subject to desorption, change pH of the medium and favorably affect (or not affect) secretion processes and biocenose of the gut organisms. Such natural enterosorbents or dietary fibers as, e.g., polysaccharides (cellulose, agarose, pectin, chitin, chitosan etc.) are the most suitable for the prophylactics of human organism intoxication with heavy metals [249, 250, 253-256].

One of the challenging polysaccharides which are systematically used for lead intoxication prophylactics is pectin [257, 258]. Free carboxyl groups of galacturonic acid set conditions for pectin acid ability to bind the ions of heavy metals in the gastrointestinal tract (GIT), this process being followed by subsequent formation of insoluble complexes (pectinates, pectates) which are not absorbed in the bowels and are led out of the organism [259]. This property is used for prophylactics of intoxication with the salts of heavy metals. In addition, pectin is a colloidal substance with pronounced sorption properties. When in the GIT, pectin forms gels. These physical-chemical peculiarities are likely to be responsible for its protective action in case of intoxications. In the process of swelling pectin dehydrates the digestive channel and passing along the bowels captures toxic substances. Pectin is not digested until it reaches the colon. It combines with some heavy metals thus forming insoluble salts which are not absorbed by the mucous membrane of the gastrointestinal channel and are led out of the organism. It was found in [257, 258], that the unit weight and the degree of etherification, i.e. concentration of free methoxyl pectin groups regulate their sensitivity and activity in complex formation. Low-etherified pectin easily forms pectinates of metals, lead included, whereas high-etherified (methoxylated) pectin obduces the intestinal wall and, owing to the gel-filtration mechanism, lowers absorption of heavy metals. The results of the experiments performed in [257, 259, 260] demonstrate that the most favorable effect on chelate formation, which is expressed in maximal lead excretion with urine and feces and a decrease of lead content in bones, is reached when the degree of pectin etherification is 50—60%. Long-term ingestion of pectin did not cause any complications or side effects [258].

Nowadays there is no unified technique of studying sorption of heavy metals, in particular lead and cadmium, with vegetable enterosorbents in model conditions [258, 261]. The authors of [262-265268-271] described the technique intended for evaluating BAS ability to sorb lead and cadmium from the solutions of their salts *in vitro*. At that the conditions (pH of the medium, mixing intensity and duration) were analogous to those in the stomach and bowels of a man. The objects of investigation were dietary fibers, i.e. polysaccharides (amorphous wood cellulose, apple pectin, potato starch, inulin extracted from chicory roots and chitin from hiratake carposome (*Pleurotus ostreatus*).The concentration of lead and cadmium cations in solid BAS residues was determined by the authors of [262-265] using atomic-emission and atomic-absorption spectroscopy. The error in determining the metal mass was 10%. A relative standard deviation during the evaluation was 0.05.

A series of polysaccharide sorption power is similar for Pb^{2+} and Cd^{2+} and is as follows: cellulose > pectin > starch > inulin [265]. It was shown in [262-265] that there is a correlation between the weight of the solid residue of the product made of vegetable raw and its ability to sorb lead and cadmium from the solutions of their salts. The larger the solid residue weight, the higher the sorption power of the product. In addition, a decrease in BAS particle sizes yields an increase in the sorption power. Interesting results were obtained in [262-265] when studying joint sorption of Pb^{2+} and Cd^{2}. An antagonistic effect of the metals when Pb^{2+} sorption remained unchanged whereas Cd^{2+} sorption substantially lowered was demonstrated.

The authors of [266] studied sorption of cesium, strontium and cadmium by dry and live biomass of hiratake and the chitin extracted from it (Table 26). It is apparent from Table 26 that sorption in the Cs < Sr < Cd < Pb series increases. It is specified in literature [83, 267, 268] that chitin binds ions of metals in the result of complex formation, ion exchange or surface sorption since there are carboxyl, hydroxyl and acetamide groups in its macromolecule. Chitin ability for complex formation is attributed to a high electrodonor ability of nitrogen and oxygen atoms.

Polysaccharides sorb well not only heavy metals, but also other substances. For example, cellulose adsorbs phenol and toluene from a gas phase and water solutions [269], cellulose acetates with different degrees of displacement sorb gaseous ammonia [270]. Chitosan can serve as a carrier of pharmaceutical compositions [271]. Sorbents can be produced from an enzymatically modified flax fiber [272].

Table 26. Averaged mass fraction values of metals (ω_{aver}) sorbed by dry and live biomass of hiratake and chitin extracted from it [266]

Salt	$CsNO_3$	$Sr(NO_3)_2$	$CdSO_4$	$Pb(NO_3)_2$
Dry hiratake biomass				
ω_{aver}, mass%	24±2	40±2	40±4	76±7
Live hiratake biomass				
ω_{aver}, mass%	22±2	25±2	27±3	43±3
Chitin from hiratake carposome				
ω_{aver}, mass%	19±1	44±1	81±1	84±2

8. Novel Nano-Polymers for Medical Applications

Nano polymers are molecules with size less than 100 nm with molecular weight range from 1-10000 kDa. Nano-polymers can be soluble or non-soluble. Injection of particularly high molecular weight (>1000 kDa) nano-polymers into the blood stream enhances blood flow and tissue perfusion. These molecules do not produce toxic effects on cells and their lifespan in the circulation can be controlled. Because of their water and blood soluble properties such nano-polymers as Polyethylene glycol (PEG), Polymethylmetacrylate (PMMA), Polyacrylamide (PAA) and Chitosan/Chitooligosaccharide (COSs) have gained interest [273-286]. The latter property permits nano-polymers integration into systemic circulation and affects bio-distribution and properties of drugs they carry [278, 279, 282 287-302]. Nano-medical technology will achieve significant advances in areas such as nano-devices, pharmacological therapeutics, diagnostic, and regenerative medicine [274-284, 286, 287, 289, 292-295, 297, 299, 301-316]. With nanotechnology-based drug delivery systems, important improvement in pharmacokinetics of drugs will take place due to subsequent increase of bioavailability. Nanotechnology in modern pharmacology will help to provide more accurate control of dosing and significantly decrease drug toxicity [291, 317-321]. Improvement in pharmacodynamic parameters will occur due to increased drug concentration in target tissues due to advanced perfusion enhancement [278, 279, 282, 288, 292, 294-300, 306, 307, 313, 314, 322, 323].

Nanotechnology offers a broad range of opportunities for improving diagnostic and therapeutic interventions in medicine. Viscoelastic and carrier properties of the applied polymers and composits were reported as factors in the improved survival. Over the years, blood-soluble nano-polymers have been shown to produce positive hemodynamic effects when intravenously administered in various acute and chronic animal models [276, 278, 282, 288, 295-297, 299, 307, 313, 314]. Nano-polymers cause an increase in circulation a decrease in both blood pressure and peripheral vascular resistance [278, 279, 282, 288, 290, 291, 295-300, 302, 306, 307, 313-318, 320, 322, 324, 325]. Nano-polymers significantly increase collateral blood flow in rabbits and the number of functional capillaries in normal and diabetic rats [279, 282, 291, 298, 300, 302, 307, 313, 314, 322, 324]. In these experiments, nano-polymers were injected at concentrations 50–100 times higher than minimal effective dose with no demonstrable acute toxicity [278, 282, 297]. The mechanisms by which nano-compounds comprised of a PEG and co-polymers appear to have major clinical relevance [278, 279, 282, 287, 290, 292-295, 297, 298, 300, 301, 307, 313, 314, 322]. PEG is well known and characterized in polymer chemistry. Its properties in the water-protein solution and blood have been thoroughly investigated. It is one of the most widely used non-toxic high molecular weight polymers in medical nano-technology and has known properties to benefit vascular circulation [279, 293, 298, 300, 302, 307, 322, 324, 325]. The following Table 27 represents a compendium of selected polymers and its copolymers which have been used for broadly various clinical applications.

Table 27. Nano-component prototypes of selected polymers and its copolymers for clinical applications

Polymer Name	Molecule Type	Function	Application
1	2	3	4
Poly(ethylene oxide)-b-poly (ε-caprolactone)	Polymeric micellar	Reduce kidney uptake, increased levels in blood	Cyclosporine A (CyA) carrier with low toxicity
Poly-ε-caprolactone (PECL)	Nanocapsules	Colloidal drug carrier/Transcellular penetration	Systems loaded with a fluorescent dye
Poly(ethylene glycol) (PEG)-coated PECL	Nanocapsules	Colloidal drug carrier/Transcellular penetration	Systems loaded with a fluorescent dye
Polyethylene glycol and poly-lactic/poly-glycolic acid co-polymer	Nanocapsules/Polymer nano-carriers (PNC)	Intravascular delivery systems	Vascular targeting for ischemia treatment
Poly(ethylene) glycol	Nano-polymeric carrier	Nano-sized drug delivery in blood/plasma	AIDS drugs cellular modification, novel pharmacological properties

Polymer Name	Molecule Type	Function	Application
1	2	3	4
	Blood-soluble drag reducing polymers (DRPs)	Enhance hemodynamics	Treatment of hemorrhagic shock
Methoxy poly(ethylene glycol)-b-poly(caprolactone) co-polymers	Nano-sized micelles	Drug delivery in blood/plasma	Increase of drug kinetic
Poly(ethylene oxide)	Blood-soluble drag-reducing polymer (DRP)	Preservation of tissue perfusion and oxygenation	Treatment of hemorrhagic shock
	Drag-reducing polymers (DRPs)/soluble linear macromolecules	Modification of microvascular resistance through an increase in capillary volume	Vascular targeting for ischemia treatment
Chitosan-poly(ethylene oxide) (PEO) co-polymer	Nanofibers of chitosan and polyethylene oxide (PEO)	Matrix for nanodevises	Nano-chips and delivery systems
Chitosan/poly(ethylen-oxide-alumina) co-polymer	Prototype polymer carriers (spheres, elliptical disks)	Carriers for intracellular drug delivery	Intraendothelial drug delivery
Poly(methyl methacrylate) and poly(ethylene oxide) co-polymer	Surface modified liposomes	Increased antitumor activities	Doxorubicin (DOX)-loaded liposomes
Polysaccharide	Drag-reducing polymers (DRPs)/soluble linear macromolecules	Prolong survival time following acute myocardial ischemia (AMI)	Treatment and prevention of tissue hypoperfusion
Chitosan	Chitosan nanoparticles	Gastrointestinal tract cell transfection	Human insulin gene non-viral vector
	Soluble linear macromolecules	Vaccine enhancement/delivery platform for the sustained, local delivery	New method of vaccination
	Particle/coating	Protein Carrier across the Blood Brain Barrier to target cerebral amyloid	Alzheimer's disease (AD) treatment
Chitosan (CS)-poly-ε-caprolactone (CS)-coated PECL	Nanocapsules	Colloidal drug carrier/Transcellular penetration	Systems loaded with a fluorescent dye
Chitooligosaccharide	Water-soluble compound	Facilitated access to systemic circulation	Enhanced blood drug distribution
Polysaccharide carbohydrate $(C_6H_{10}O_5)n$-Fe	Polymer-coated magnetic nanoparticles (starch-coated)	Delivery of hyperthermic thermoseeds	Magnetic targeted hyperthermia
Chitosan-Fe	Polymer-coated magnetic nanoparticles (chitosan-coated, and starch-coated)	Electro-msgnretic activation and precise drug-cell targeting	Targeted hyperthermia

Table 27. (Continued)

Polymer Name	Molecule Type	Function	Application
1	2	3	4
Chitin-alginate co-polymer	Water-soluble chitin-alginate-PEI-DNA nanoparticles	Nonwoven fibrous scaffold vehicle for DNA delivery	Matrix capable of delivering genes to direct and support cellular development in tissue engineering
Photocrosslinkable chitosan	Soluble chitosan hydrogel	Prevention of drug inactivation due to heat, proteolysis and low pH	Prolong biological half-life time of FGF-2
Chitin (Poly-N-acetyl glucosamine) deacetylate	Nano carrier	Systemic hemostatic carrier	Hemorrhage treatment
Poly (lactide-co-glycolide)-chitosan	Nanospheres	Nonviral vectors	Synthetic vectors for gene therapy
N-octyl-O-sulfate chitosan	Micelles	Drug carrier/reduce toxicity and improve bioavailability	New formulation of Paclitaxel (Taxol)-PTX with better efficacy and fewer side effects for cancer treatment
	Micelles	Increase water/lipids solubility of medical components	New drug delivery systems
Poly(lactic acid)	Nanoparticles	Increase Hydrophilicity	Facilitate brain circulation
Poly(DL-lactide-co-glycolide)	Nanoparticles (NP)/nanocapsules (NC)	Sustained release polymeric gene delivery system	Timed release of encapsulated pDNA. Alternative to viral vectors
Poly(methyl methacrylate)	Polymer films	Reduction in plasma protein adsorption, and platelet adhesion	Control of thrombus formation in cardiovascular applications

Nano-polymers are attractive materials for both biomedical engineering and medical applications. Rational design of nano-polymer carrier geometry will help optimize customized tissue and cell targeted therapeutics. Further understanding of how nano-polymers interact both *in vivo* and *in vitro* with different cells will open new horizons in nano-medicine. It is easily foreseeable that designer nano-polymers can be tailored to induce favorable cell response, in order to precisely intervene in disease mechanism.

CONCLUSION

The performed analysis of the literary data allows making a conclusion that a complex investigation of interrelation between the structure, thermodynamic characteristics and sorption properties of polysaccharides makes it possible to purposefully select effective means of BAS supply to a human organism.

ACKNOWLEDGMENTS

The work was performed with financial support of the:

1. Russian Foundation Basic Research (grant No 08-03-97052 r_povolzhie_a).
2. Ministry of Education and Science of Russia in framework of the Analytical departmental goal program "Development of scientific potential of higher school (2009-2010)", (project No 2.1.2/1056).

REFERENCES

[1] Rabinovich, I.B., Martynenko, L.Ya., Sheiman, M.S., Ovchinnikov, Yu.V., Karyakin, N.V., and Zarudaeva, S.S., Thermodynamic investigations of physical-chemical nature of polymer-plasticizer systems, *Tr. po Khimii i Khim. Tekhnologii,* Gorki: GSUniv. Press, 1972, no. 2, pp. 98-111 (in Russian).

[2] Rabinovich, I.B., Mochalov, A.N., Tsvetkova, L.Ya., Khlystova, T.B., Moseyeva, Ye.M., and Maslova, V.A., Calorimetric methods and determination results of compatibility of a number of plasticizers with polymers, *Acta Polym.,* 1983, Bd. 34, H. 8, S. 482-488.

[3] Uryash, V.F., Rabinovich, I.B., Mochalov, A.N., and Khlyustova, T.B. Thermal and calorimetric analysis of cellulose, its derivatives and mixtures with plasticizes, *Thermochim. Acta.,* 1985, vol. 93, pp. 409-412.

[4] Uryash, V.F., Chemical thermodynamics of biologically active substances and processes with their participation, Doctoral (Chem.) Dissertation, Nizhny Novgorod: NII khimii NNGU, 2005 (in Russian).

[5] Uryash, V.F., Thermodynamics of chitin and chitosan. In. Chitin and Chitosan. Production, Properties and Usage, Skryabin, K.G., Vikhoreva, G.A., and Varlamov, V.P., Eds., Moscow*: Nauka*, 2002, pp.119-129 (in Russian).

[6] Biochemical Thermodynamics, Jones, V.N., Ed., Amsterdam: Elsevier Publ. Co., 1979.

[7] Kalous, V. and Pavlicek, Z., Biofyzikalni Chemie, Praha: Nakladatestvi Techn. Literat, 1985 (in Czech).

[8] Mrevlishvili, G.M., Low-Temperature Calorimetry of Biological Macromolecules, Tbilisi: Metsniereba Publ., 1984 (in Georgian).

[9] Physics and Chemistry of the Organic Solid State, Fox, D., Labes, M.M., and Weissberger, A., Eds., NY, London: John WileyandSons Publ., 1965, vol.1, pp. 9-160.

[10] Stull, D.R., Westrum, E.F., and Sinke, G.C., Chemical Thermodynamics of Organic Compounds, NY: J.Wiley and Sons. Inc. Publ., 1969.

[11] Hutchens, J.O., Cole, A.G., and Stout, J.W., Heat capacities from 11 to 305 K and entropies of hydrated and anhydrous bovine zinc insulin and bovine chymotrypsinogen A, *J. Biol. Chem.,* 1969, vol. 244, no. 1, pp. 26-32.
[12] Rabinovich, I.B., Nistratov, V.P., Telnoy, V.I., and Sheiman, M.S., Thermodynamics of Organometallic Compounds, Nizhny Novgorod: NNUniv. Press, 1996 (in Russian).
[13] Karyakin, N.V., Thermodynamics of Aromatic Heterochain and Heterocyclochain Polymers, Nizhny Novgorod: NNUniv. Press, 1998 (in Russian).
[14] Lebedev, B.V., *Thermodynamics of Polymers*, Nizhny Novgorod: NNUniv. Press, 1989 (in Russian).
[15] Van Krevelen, D.W., *Properties of Polymers Correlations with Chemical Structure*, Amsterdam: Elsevier Publ. Co., 1972.
[16] *Water in Polymers*, Rowland, S.P., Ed., Washington, D.C.: Amer. Chem. Soc. Publ., 1980.
[17] Mochalov, A.N., Khlyustova, T.B., Ioelovich, M.Ya., and Kaimin', I.F., The effect of cellulose crystallinity degree on its heat capacity, *Khimiya Drevisiny,* 1982, no. 4, pp. 66-68 (in Latvian).
[18] Mikhailov, N.V. and Fainberg, E.Z., Heat capacity and phase state of cellulose fibers with different structure, *Vysokomelek. Soedin.,* 1962, vol. 4, no. 2, pp. 230-236 (in Russian).
[19] Fainberg, E.Z. and Mikhailov, N.V., Investigation of temperature dependence of cellulose fiber heat capacity, *Vysokomolek. Soedin.*, 1967, vol. A9, no. 4, pp. 920-926 (in Russian).
[20] Kozlov, N.A., Rabinovich, I.B., Medved', Z.N., Sheiman, M.S., and Mochalov, A.N., Low-temperature heat capacity of cellulose of different dehydration methods, *Cellulose Chem. Technol.*, 1972, vol. 6, no. 6, pp. 601-607.
[21] Karachevtsev, V.G. and Kozlov, N.A., Investigation of cellulose thermodynamic properties at low temperatures, *Vysokomolek. Soedin.*, 1974, vol. A16, no. 8, pp.1892-1897 (in Russian).
[22] Kaimin', I.F., Investigation of cellulose heat capacity, *Vysokomolek. Soedin.*, 1979, vol. B21, no. 5, pp. 331-334 (in Russian).
[23] Hatakeyama, T., Nakamura, H., and Hatakeyama, H., Studies on capacity of cellulose and lignin by differential skanning calorimetry, *Polymer,* 1982, vol. 23, no. 12, pp. 1801-1804.
[24] Ioelovich, M.Ya., Inhomogeneity of cellulose supramolecular structure and properties, Extended Abstract of Doctoral (Chem.) Dissertation, Riga: Institut Khimii Drevesiny LatAN, 1991 (in Latvian).
[25] Wunderlich, B. and Baur, H., Heat Capacities of Linear High Polymers, Heidelberg, NY: Springer Publ., 1970.
[26] Godovskii, Yu.K., Thermal Physical Methods of Polymer Investigation, Moscow: Nauka, 1976 (in Russian).
[27] Kobeko, P.P., Amorphous Substances: Physical-Chemical Properties of Elementary and High-Molecular Amorphous Substances, Moscow: AN SSSR Publ., 1952 (in Russian).
[28] Uryash, V.F., Mochalov, A.N., Kupriyanov, V.F., Smirnov, A.G., Kuleshova, T.M., and Samoshkin, V.I., Heat capacity and thermodynamic functions of glycerin trinitrate, *Zhurn. Obshchei Khimii*, 1997, vol. 67, no. 4, pp. 588-591 (in Russian).

[29] Uryash, V.F., Kupriyanov, V.F., Kokurina, N.Yu., Smirnov, A.G., and Kuleshova, T.M., Heat capacity and thermodynamic functions of diethylene glycoldinitrate, *Zhurn. Obshchei Khimii*, 2000, vol. 70, no. 5, pp. 719-721 (in Russian).

[30] Kelley, K.K., Parks, G.S., and Huffman, H.M. A New method for extrapolating specific heat curves of organic compounds below the temperatures of liquid air, *J. Phys. Chem.,* 1929, vol., 33, no. 11, pp.1802-1805.

[31] Mochalov, A.N., Rabinovich, I.B., Uryash, V.F., Khlyustova, T.B., and Mikhailov, B.I., Heat capacity, physical transitions and structure of nitrocellulose, *Thermodynamics of organic substances*, Gorki: GSUniv. Press, 1981, pp. 16-19 (in Russian).

[32] Kaimin', I.F. and Ioelovich, M.Ya., Dilatometry and thermomechanics methods in investigation of cellulose and its derivatives. In: Methods of Cellulose Investigation, Karlivana, V.P., Ed., Riga: Zinatne Publ., 1981, pp. 73-95 (in Latvian).

[33] Kargin, V.A., Kozlov, P.V., and Van Nai Chan, Cellulose vitrification temperature, *Dokl. AN SSSR*, 1960, vol.130, no. 2, pp. 356-358 (in Russian).

[34] Mikhailov, N.V. and Fainberg, E.Z., Cellulose phase state in oriented fibers, *Dokl. AN SSSR,* 1956, vol. 109, no. 6, pp. 1160-1162 (in Russian).

[35] Ioelovich, M.Ya., Kaimin', I.F., and Veveris, G.P., Process of amorphous cellulose crystallization, *Vysokomolek. Soedin.*, 1982, vol. A24, no. 6, pp. 1224-1228 (in Russian).

[36] Klason, C. and Kubat, J., Thermal transition in cellulose, *Svenk Papperstidn.,* 1976, vol. 79, no. 15, pp. 494-500.

[37] Nakamura, S., Gillham, J.K., and Tobolsky, A.V., Torsional braid analyses of cellulose, *Rep. Progr. Polymer Phys. Jap,.* 1970, vol.13, pp. 89-90.

[38] Zelenev, Yu.V. and Glazkov, V.I., Relaxation processes in cellulose and its derivatives *Vysokomolek. Soedin.*, 1972, vol. A14, no. 1, pp.16-22 (in Russian).

[39] Naimark, N.I. and Fomenko, B.A., Cellulose vitrification, *Vysokomolek. Soedin.*, 1971, vol. B13, no. 1, pp. 45-46 (in Russian).

[40] Bradly, S.A. and Carr, S.H., Mechanical loss processes in polysaccharides, *J. Polym. Sci.* A-2, 1976, vol.14, no. 1, pp. 111-124.

[41] Mikhailov, G.P., Artyukhov, A.I., and Borisova, T.I., Peculiarities of cellulose hydroxyl group relaxation at low temperatures, *Vysokomolek. Soedin.*, 1967, vol. B9, no. 2, pp.138-141 (in Russian).

[42] Klarmen, A.F., Galanti, A.V., and Sperling, L.H. Transition temperatures and structural correlation for cellulose trimesters, *J. Polym. Sci.* A-2, 1969, vol. 7, no. 7, pp.1513-1523.

[43] Borisova, T.I., Dielectric method of cellulose investigation. In: Methods of Cellulose Investigation, Karlivana, V.P., Ed., Riga: Zinatne Publ., 1981, pp. 96-110 (in Latvian).

[44] Cellulose and Cellulose Derivatives. Bikales, N.M. and Segal, L., Eds., NY: John WileyandSons, Inc. Publ., 1971, Part IV.

[45] Zhbankov, R.G. and Kozlov, P.V., Physics of Cellulose and its Derivatives, Minsk: Nauka i Tekhn. Publ., 1983 (in Byelorussian).

[46] Tarchevskii, I.A. and Marchenko, G.N., Cellulose Biosynthesis and Structure, Moscow: Nauka, 1985 (in Russian).

[47] Uryash, V.F., Faminskays, L.A., Kokurina, N.Yu., Gruzdeva, A.E., Grishatova, N.V., and Larina, V.N., Thermodynamic analysis and calorimetry of microcrystalline cellulose (MCC) in the 80-550 K range, *13-th All-Russian Conference on Thermal*

Analysis (and Calorimetry): Proceedings, Samara: SamGASA Publ., 2003, pp.104-106 (in Russian).

[48] Larina, V.N., Uryash, V.F., Kokurina, N.Yu., and Novoselova, N.V., Influence of degree of order on thermochemical characteristics of cellulose and solubility of water in it, *17-th International Conference on Chemical Thermodynamics in Russia: Proceedings,* Kazan: KSTU Publ., 2009, vol.1, p.100 (in Russian).

[49] Uryash, V.F., Larina, V.N., Kokurina, N.Yu., and Novoselova, N.V., Thermochemical characteristics of cellulose and its mixtures with water, *Zhurn. Fiz. Khimii,* 2010, vol. 84 (in press) (in Russian).

[50] Ioelovich, M.Ya., Enthalpy of cellulose formation and solubilization. *12-th All-Russian Conference on Chemical Thermodynamics and Calorimetry: Proceedings,* Gorki: GSUniv. Press, 1988, part 2, pp.189-191 (in Russian).

[51] Matyushin, Yu.N., Korchatova, L.I., Sopin, V.F., Marchenko, G.N., and Lebedev, Yu.A., Investigation of the effect of cellulose drying conditions on its thermal value. *1-st All-Russia Conference "Cellulose synthesis and its control": Proceedings, Kazan:* KKhTI Publ., 1980, pp. 36-37 (in Russian).

[52] Matyushin, Yu.N., and Lebedev, Yu.A., Calorimetric methods of cellulose investigations. *All-Russian conference "Methods of cellulose investigation": Proceedings,* Riga: Institut Khimii Drevesiny LatAN Publ., 1988, pp. 135-138 (in Latvian).

[53] Prokhorov, A.V., Calculation of thermodynamic characteristics of some cellulose modifications for predicting its processing conditions, *Khimiya Drevisiny,* 1981, no. 4, pp. 73-80 (in Latvian).

[54] Nikitin, N.I., Chemistry of Cellulose and Wood, Moscow: AN SSSR Publ., 1962 (in Russian).

[55] Zharkovskii, D.V., Physical-Chemical Investigations of Cellulose and its Ethers, Minsk: Bel. Inst. Mekh. Sel. Khoz. Publ., 1960 (in Byelorussian).

[56] Papkov, S.P., and Fainberg, E.Z., Interaction of Cellulose and Cellulose Materials with Water, Moscow: Khimiya, 1976 (in Russian).

[57] Tsvetkov, V.G., Kaimin', I.F., Ioelovich, M.Ya., and Rabinovich, I.B., Enthalpy of water interaction with cellulose of different degree of crystallinity, *All-Russia Seminar "Crystallization of polysaccharides and their interaction with water: Proceedings,* Riga: Zinatne Publ., 1979, pp. 6-8 (in Latvian).

[58] Tsvetkov, V.G., Ioelovich, M.Ya., Kaimin', I.F., and Reizin'sh, R.E., Enthalpy of interaction of cellulose with different degree of crystallinity with water, *Khimiya Drevisiny,* 1980, no. 5, pp. 12-15 (in Latvian).

[59] Osovskaya, I.I., and Mishchenko, K.P., Wetting heats of different cellulose materials *All-Russia Seminar "Crystallization of polysaccharides and their interaction with water: Proceedings,* Riga: Zinatne Publ., 1979, pp. 9-12 (in Latvia).

[60] Tsvetkov, V.G., Kaimin', I.F., Ioelovich, M.Ya., and Prokhorov, A.V., Enthalpy of interaction of cellulose and some of its model compounds with solvents, *Thermodynamics of organic compounds,* Gorki: GSUniv. Press., 1982, pp. 54-60 (in Russian).

[61] Tsvetkov, V.G., Enthalpy of mixing of cellulose and its derivatives with low-molecular substances. In: Methods of Cellulose Investigation, Karlivina, V.P., Ed., Riga: Zinatne Publ., 1981, pp. 126-137 (in Latvian).

[62] Marchenko, G.N. and Tsvetkov, V.G., Solvation and thermochemistry of nonaqueous cellulose solutions. Thermochemistry of cellulose nitrate ethers formation. In: Physical-Chemical Foundations and Hardware Implementation of the Technology of Pyroxylin Gunpowder Manufacturing. Cellulose Nitrates, Marchenko, G.N., Ed., Kazan: FEN Publ., 2000, vol. 1, pp. 196-282 (in Russian).

[63] Klenkova, N.I., Cellulose Structure and Relaxation Ability, Leningrad: Nauka, 1979 (in Russian).

[64] Ioelovich, M.Ya. and Kreitus, A.E., Analytical description of cellulose hydrophilic properties, *Khimiya Drevisiny,* 1983, no. 3, pp. 3-6 (in Latvian).

[65] Jenckel, E., Gorke, K., Zur Kalorimetrie der Hochpolymeren. II. Die integralen Verdünnungswärmen der Lösungen des polystyrols mit Äthylbenzol, Toluol, Chlorbenzol und Cyclohexan, *Z. Elektrochemie,* 1956, Bd. 60, H. 6, S. 579-587.

[66] Meerson, S.I. and Lipatov, S.M., Thermochemical methods of polymer investigation, *Zhurn. Vses. Khim. Obshchestva im. D.I. Mendeleeva*, 1961, vol. 6, no. 4, pp. 412-416 (in Russian).

[67] Meerson, S.I., Determination of polymer salvation degree using calorimetric technique, *Kolloidn. Zhurn.*, 1969, vol. 31, no. 3, pp. 421-426 (in Russian).

[68] Tsvetkov, V.G. and Tsvetkova, L.Ya., Final differential heats of electrolyte solubilization in water solutions, *Zhurn. Fizich. Khimii*, 1974, vol. 53, no. 7, pp. 1822-1824 (in Russian).

[69] Tager, A.A., Thermodynamics of concentrated polymer solutions, *Vysokomolek. Soedin.*, 1971, vol. A13, no. 2, pp. 467-484 (in Russian).

[70] Nikolaev, P.N. and Rabinovich, I.B., Thermochemistry of isotope compounds. 1. Effect of hydrogen displacement with double-weight hydrogen on benzole heat capacity. *Tr. po Khimii i Khim. Tekhnologii,* Gorki: GSUniv. Press., 1961, no. 2, pp. 242-250 (in Russian).

[71] Skuratov, S.M., Heat capacity of the solvent related to high-polymer substances, *Kolloidn. Zhurn.*, 1947, vol. 9, no. 2, pp. 133-140 (in Russian).

[72] Uryash, V.F., Kokurina, N.Yu., Maslova, V.A., Larina, V.N., and Iosilevich, I.N., Calorimetric investigations of fungic chitin and its mixtures with water, *Vestnik of Lobachevsky State University of Nizhni Novgorod. Seriya Khimiya,* Nizhny Novgorod: NNUniv. Press, 1998, vyp. 1, pp. 165-170 (in Russian).

[73] Uryash, V.F., Karyakin, N.V., and Gruzdeva, A.E., Optimization of the production process of biologically active substances based on calorimetric data, *Perspektivnye Materialy*, 2001, no. 6, pp. 61-69 (in Russian).

[74] Tsvetkova, L.Ya., Novoselova, N.V., Golitsin, V.P., Ivanov, A.V., Khlyustova, T.B., and Uryash, V.F., Thermodynamic characteristics of chitin and chitosan, *Zhurn. Khimich. Termodinamiki i Termokhimii*, 1993, vol. 2, no. 1, pp. 88-93 (in Russian).

[75] Plisko, E.A., Nud'ga, L.A., and Danilov, S.N., Chitin and its chemical transformations, *Uspekhi Khimii*, 1977, vol. 46, no. 8, pp. 1470-1487 (in Russian).

[76] Vikhoreva, G.A., Gorbacheva, I.N., and Gal'braikh, L.S., Chemical modification of hydrobionite polysaccharides, *Khimich. Volokna*, 1994, no. 5, pp. 37-45 (in Russian).

[77] Gal'braikh, L.S., Chitin and chitosan: structure, properties, application, *Sorosovskii Obrazovatel'nyi Zhurnal*, 2001, vol. 7, no. 1, pp. 51-56 (in Russian).

[78] Novoselov, N.P. and Sashina, E.S., Modern ideas on cellulose, chitin and chitosam structure. Mechanism of their solubilization and biological activity. In: Biologically

Active Substances in Solutions: Structure, Thermodynamics, Reactivity, Kutepova, A.M., Ed., Moscow: Nauka, 2001, pp. 336-397 (in Russian).
[79] Genin, Ya.V., Sklyar, A.M., Tsvankin, D.Ya., Gamzazade, A.K., Rogozhin, S.V., and Pavlova, S.A., X-ray investigations of chitosan films, *Vysokomolek. Soedin.*, 1984, vol. A26, no. 11, pp. 2411-2416 (in Russian).
[80] Elagin, A.A., and Pertsin, A.N., Conformation of isolated chitin macromolecule, *Vysokomolek. Soedin.*, 1983, vol. A25, no. 4, pp. 804-811 (in Russian).
[81] Mogilevskaya, E.L., Akopova, T.A., Zelenetskii, A.N., and Ozerin, A.N., Crystalline structure of chitin and chitosan, *Vysokomolek. Soedin.*, 2006, vol. A48, no. 2, pp. 216-226 (in Russian).
[82] Mar'in, A.P., Feofilova, E.P., Genin, Ya.V., Shlyapnikov, Yu.A., and Pisarevskaya, I.V., Effect of crystallinity of sorption and thermal properties of chitin and chitosan, *Vysokomolek. Soedin.*, 1982, vol. Б24, no. 9, pp. 658-662 (in Russian).
[83] Gorovoi, L.F. and Kosyakov, V.N., Sorption properties of chitin and its derivatives. In: Chitin and Chitosan. Production, Properties and Usage, Skryabin, K.G., Vikhoreva, G.A., Varlamov, V.P., Eds., Moscow: Nauka, 2002, pp. 217-246 (in Russian).
[84] Uryash, V.F., Kokurina, N.Yu., Lariana, V.N., Varlamov, V.P., Il'ina, A.V., Grishatova, N.V., and Gruzdeva, A.E., Effect of acid hydrolysis on heat capacity and physical transitions of chitin and chitosan, *Vestnik of Lobachevsky State University of Nizhni Novgorod,* Nizhny Novgorod: NNUniv. Press, 2007, no. 3, pp. 98-104 (in Russian).
[85] Khlyustova, T.B., Thermodynamics and physical-chemical analysis of cellulose nitrate and acetate mixtures with esters (plasticizers). Extended Abstract of Candidate (Chemistry) Dissertation (PhD), Leningrad: Len. Tekhnologich. Inst. Tselyulozno-Bumazhnoi Prom., 1986 (in Russian).
[86] Mochalov, A.N., Khlyustova, T.B., and Malinin, L.N., Heat capacity and physical transitions of cellulose acetates with different substitution, *Fiziko-khimicheskiye osnovy sinteza i pererabotki polimerov,* Gorki: GSUniv. Press, 1982, pp. 59-63 (in Russian).
[87] Rabinovich, I.B., Zarudaeva, S.S., Mochalov, A.N., Luk'yanova, N.V., Pegova, E.B., Khlyustova, T.B., Uryash, V.F., Diagram of physical states of nitrocellulose (NC)-nitroglycerin (NG) system, *Tr. po Khimii i Khim. Tekhnologii,* Gorki: GSUniv. Press, 1974, no. 1, pp. 118-122 (in Russian).
[88] Uryash, V.F., Mochalov, A.N., and Kupriyanov, V.F., Heat capacity and physical-chemical analysis of high-nitrogenous cellulose nitrate and its mixtures with dimethyl phthalate, *Fizika protsessov sinteza i svoistva polimerov,* Gorki: GSUniv. Press, 1988, pp. 47-51 (in Russian).
[89] Nakamura, K. and Ookawa, T., Studies on viscoelasticiti of cellulose derivatives films. III. Second-order transition points of cellulose nitrates, *Chem. High Polym. Japan*, 1957, vol. 14, no. 150, pp. 544-550.
[90] Bakaev, A.S., Ul'yanov, V.P., Shneerson, R.I., and Papkov, S.P., Structural transitions in nitrocellulose - dibutyl phthalate and nitrocellulose - castor oil structures, *Tr. MChTI im. D. I. Mendeleeva,* Moscow: MKhTI Publ., 1970, vyp. 66, pp. 219-223 (in Russian).
[91] Sorokin, G.A., Thermographic determination of the content of polymer-bound solvent, *Vysokomolek. Soedin.*, 1971, vol. A13, no. 3, pp. 608-612 (in Russian).

[92] Timofeeva, V.G. and Kozlov, P.V., Effect of the depth of supramolecular structure dissociating on vitrification temperatures of some cellulose ethers. In: Cellulose and its Derivatives, Moscow: AN SSSR Publ, 1963, pp. 167-173 (in Russian).

[93] Kozlov, P.V., Timofeeva, V.G., and Kargin, V.A., Effect of low-molecular substances sorbed by supramolecular structures on mechanical properties of rigid polymers, *Dokl. AN SSSR*, 1963, vol. 148, no. 4, pp. 886-889 (in Russian).

[94] Klarmen, A.F., Galanti, A.V., and Sperling, L.H., Transition temperatures and structural correlation for cellulose trimesters, *J. Polym. Sci.,* A-2, 1969, vol. 7, no. 7, pp. 1513-1523.

[95] Uberreiter, K., Uber Cellulose und ihre Derivate als Flussigkeiten mit Fixierten Struktur, *Z. Phys. Chem.,* 1941, Abt.B, Bd. 48, H. 3, S. 197-218.

[96] Aziz, K. and Shinouda, H.G., Umwandlung Zweiter Ordnung in Cellulosederivaten, *Faserforsch. u. Textiltechn.,* 1973, Bd. 24, H. 12, S. 510-512.

[97] Fujimoto, T. and Inoue, Y., Glass transition of cellulose derivatives, *Chem. High Polym. Japan,* 1960, vol. 17, no. 183, pp. 436-440.

[98] Clash, R.F. and Rynkiewicz, L.M., Thermal expansion properties of plastic materials, *Ind. Eng. Chem.,* 1944, vol. 36, no. 3, pp. 279-282.

[99] Wiley, F.E., Transition temperature and cubical expansion of plastic materials, *Ind. Eng. Chem.,* 1942, vol. 34, no. 9, pp. 1052-1056.

[100] Cheperegin, E.A., Bakaev, A.S., Shneerson, R.I., and Stetsovskii, A.P., Dielectric properties and interaction of nitrocellulose with plasticizers, *Tr. MChTI im. D. I. Mendeleeva,* Moscow: MKhTI Publ., 1968, vyp. 57, pp. 209-213 (in Russian).

[101] Golovin V.A., Lotmentsev, Yu.M., and Shneerson, R.I., Investigation of tri-nitroglycerin compatibility with cellulose nitrate using static technique of saturated vapor pressure measurement, *Vysokomolek. Soedin.*, 1975, vol. A17, no. 10, pp. 2351-2354 (in Russian).

[102] Rabinovich, I.B., Mochalov, A.N., Zarudaeva, S.S., Khlyustova, T.B., Kuznetsov, G.A., Malinin, L.N., and Fridman, O.A., Physical-chemical analysis of cellulose diacetate mixtures with diethyl phthalate and P-514 polyether, *Vysokomolek. Soedin.*, 1979, vol. B21, no. 12, pp. 888-892 (in Russian).

[103] Khlyustova, T.B., Mochalov, A.N., and Kokurina, N.Yu., Physical-chemical analysis of nitrocellulose binary mixtures with triacetine and diethyl phthalate, *Termodinamika organ. soed.*, Gorki: GSUniv. Press, 1982, pp. 77-83 (in Russian).

[104] Khlyustova, T.B., Rabinovich, I.B., Mochalov, A.N., and Kokurina, N.Yu., Calorimetric determination of dimethyl phthalate solubility in nitrocellulose, *Termodinamika organ. soed.,* Gorki: GSUniv. Press., 1983, pp. 70-72 (in Russian).

[105] Zarudaeva, S.S., Pet'kov, V.I., and Malinin, L.N., Vitrification and component compatibility of cellulose diacetate with dimethyl phthalate, *Termodinamika organ. soed.,* Gorki: GSUniv. Press, 1983, pp. 73-75 (in Russian).

[106] Rabinovich, I.B., Khlyustova, T.B., and Mochalov, A.N., Calorimetric determination of thermochemical properties and phase diagram of cellulose nitrate mixtures with dibutyl phthalate, *Vysokomolek. Soedin.*, 1985, vol. A27, no. 3, pp. 525-531 (in Russian).

[107] Kozlov, P.V. and Russkova, E.F., Effect of low-molecular substances on cellulose deformation in a wide temperature range, *Dokl. AN SSSR*, 1954, vol. 95, no. 3, pp. 583-586 (in Russian).

[108] Borisova, T.I. and Chirkov, V.N., Investigation of solvent molecule mobility in a polymer matrix using dielectric relaxation, *Vysokomolek. Soedin.*, 1972, vol. A14, no. 9, pp. 1929-1935 (in Russian).
[109] Golovin, V.A. and Lotmentsev, Yu.M., Investigation of structure and thermodynamic parameters of component interaction in plasticized cellulose nitrates, *Vysokomolek. Soedin.*, 1981, vol. A23, no. 6, pp. 1310-1314 (in Russian).
[110] Timofeeva, V.G., Borisova, T.I., Mikhailov, G.P., and Kozlov, P.V., Effect of supramolecular structure dissociation on the temperature maximum of dielectric losses of cellulose triacetate. In: Cellulose and its Derivatives, Moscow: AN SSSR Publ., 1963, pp. 174-180 (in Russian).
[111] Timofeeva, V.G., Zaitseva, V.D., Bartenev, G.M., and Kozlov, P.V., Effect of low-molecular substances on structural-mechanical mobility of cellulose triacetate in a wide temperature range. In: Cellulose and its Derivatives, Moscow: AN SSSR Publ., 1963, pp. 181-185 (in Russian).
[112] Stetsovskii, A.P., Shidyakov, S.I., Kopytova, D.I., Tarasova, L.V., and Rogov, N.G., Dipole relaxation in nitrocellulose plasticized with dibutyl phthalate and castor oil, *Vysokomolek. Soedin.*, 1982, vol. B24, no. 2, pp. 150-154 (in Russian).
[113] Kaimin', I.F., Ozolinya, G.A., and Plisko, E.A., Investigation of chitosan temperature transitions, *Vysokomolek. Soedin.*, 1980, vol. A22, no. 1, pp. 151-156 (in Russian).
[114] Kaimin'sh, I.F., Physical-chemical properties of chitosan and possibilities of its practical application, *5-th All-Russia conference "New perspectives in studying chitin and chitosan": Proceedings,* Moscow: VNIIRO Publ., 1999, pp. 230-231 (in Russian).
[115] Marchenko, G.N., Tsvetkov, V.G., Marsheva, V.N., Tsvetkova, L.Ya., and Vasil'ev, N.K., Thermochemical investigation of chitin interaction with some solvents. *Thermodynamic properties of solutions*, Ivanovo: Khimiko-Tekhnologich. Institut Publ., 1984, pp. 37-39 (in Russian).
[116] Kokurina, N.Yu., Uryash, V.F., Larina, V.N., Faminskaya, L.A., Novoselova, N.V., Aktuganov, G.E., Melentiev, A.I., Zagorskaya, D.S., Nemtsev, S.V., Iliina, A.V., and Varlamov, V.P. Thermodynamic characteristics of chitin and chitosan depending on source of production and degree of order, *17-th International Conference on Chemical Thermodynamics in Russia: Proceedings,* Kazan: KSTU Publ, 2009, vol. 1, pp. 201 (in Russian).
[117] Hoffman, J.D., Williams, G., and Passaglia, E., Analysis of α-, β- and γ-relaxation processes in polychloride trifluoethylene and polyethylene: dielectric and mechanical properties. In: Transitions and Relaxations in Polymers, Boyer, R.F., Ed., NY: John WileyandSons, Inc. Publ., 1966, pp. 193-272.
[118] Bartenev, G.M. and Sanditov, D.S., Relaxation Processes in Vitreous Systems, Novosibirsk: Nauka, 1986 (in Russian).
[119] Perez, J., Investigation of polymer materials using mechanical spectrometry, *Vysokomolek. Soedin.*, 1998, vol. B40, no. 1. pp. 102-135 (in Russian).
[120] Ivanov, A.V., Tsvetkova, L.Ya., Gartman, O.R., Tel'noi, V.I., Novoselova, N.V., and Golitsin, V.P., Thermodynamics of the process of chitosan production in solutions, *6-th Intern. Conference on problems of salvation and complex formation in solutions*: Proceedings, Ivanovo: IGKhTA Publ., 1995, pp. L19-L20 (in Russian).
[121] Ur'yash, V.F., Kokurina, N.Yu., Larina, V.N., Varlamov, V.P., Il'ina, A.V., Faminskaya, L.A., and Novoselova, N.V., Dependence of thermodynamic and sorption

properties of chitin and chitosane on their origin and degree of order, *9-th Intern. conference "Modern perspectives in investigation of chitin and chitosan": Proceedings*, Moscow: VNIRO Publ., 2008, pp. 113-115 (in Russian).
[122] Calvet, E. and Prat, H., Microcalorimetrie. Applications Physico-Chimiques et Biologiques, Paris: Editeurs Libraires de l'Academie de Medicine, 1956.
[123] Gruzdeva, A.E., Uryash, V.F., Karyakin, N.V., Kokurina, N.Yu., and Grishatova, N.V., Heat capacity and physical-chemical analysis of agar and agarose, *Vestnik of Lobachevsky State University of Nizhni Novgorod. Ser. Khimiya,* Nizhny Novgorod; NNUniv. Press, 2000, vyp. 1(2), pp. 139-145 (in Russian).
[124] Stepanenko, B.N., Chemistry and Biochemistry of Carbohydrates (Polysaccharides), Moscow: Vyssh. Shk. Publ., 1978 (in Russian).
[125] Uryash, V.F., Gruzdeva, A.E., Kokurina, N.Yu., Grishatova, N.V., and Larina, V.N., Thermodynamic characteristics of amylase, amylopectin and starch within the 6—320 K range, *Zhurn. Fiz. Khimii*, 2004, vol. 78, no. 5, pp. 796-804 (in Russian).
[126] Lebedev, B.V., Bykova, T.A., Ryabkov, M.A., Vasilenko, N.G., and Muzafarov, A.M., Thermodynamic parameters of polymethyl phenylsiloxane in the 0-340 K range, *Zhurn. Fiz. Khimii*, 2000, vol. 74, no. 5, pp. 808-813 (in Russian).
[127] Lebedev, B.V., Smirnova, N.N., Ryabkov, M.V., Ponomarenko, S.A., Makeev, E.A., Boiko, N.I., and Shibaev, V.P., Thermodynamic properties of carboxysilane dendrimere of first generation with chain-terminal metoxyundecylenate groups in the 0-340 K range, *Vysokomolek. Soedin.*, 2001, vol. A43, no. 3, pp. 514-523 (in Russian).
[128] Loranskaya, T.I., Kabanova, I.N., and Klykova, E.V., Investigation of the effect of pectin-containing BAS on gastroduodenal motility of patients with functional dyspepsia, *Voprosy Pitaniya*, 2002, vol. 71, no. 2, pp. 31-33 (in Russian).
[129] Kochetkov, N.K., Bochkov, A.F., Dmitriev, B.A., Usov, A.I., Chizhov, O.S., and Shibaev, V.N., Chemistry of Carbohydrates, Moscow: Khimiya Publ., 1967 (in Russian).
[130] Uryash, V.F., Gruzdeva, A.E., Kokurina, N.Yu., Grishatova, N.V., and Larina, V.N., Thermodynamic characteristics of pectin of different etherification degree in the 6-330 K range, *Zhurn. Fiz. Khimii,* 2005, vol. 79, no. 8, pp. 1383-1389 (in Russian).
[131] Kretovich, V.L., Biochemistry of Plants, Moscow: Vyssh. Shk. Publ., 1986 (in Russian).
[132] Sheveleva, S.A., Probiotics, prebiotics and probiotic products. Modern state, *Voprosy Pitaniya,* 1999, vol. 68, no. 2, pp. 32-39 (in Russian).
[133] Uryash, V.F., Gruzdeva, A.E., Grishatova, N.V., Kokurina, N.Yu., Faminskaya, L.A., Larina, V.N., and Stepanova, E.A., Physical-chemical properties of inulin-polysaccharide contained in topinambour, *Non-traditional natural resources, innovative technologies and ptoducts: Sb. nauchn. tr.,* Moscow: Ros. Akademiya Estestv. Nauk Publ., 2003, vyp. 9, pp. 182-188 (in Russian).
[134] Smirnova, N.N., Lebedev, B.V., and Vunderlikh, B., Heat capacity and thermodynamic functions of poly(oxy-1,4-phenylene), poly(oxy-1,4-benzoyl) and poly(oxy-2,6-dimethyl-1,4-phenylene) in the 0-325 K range, *Vysokomolek. Soedin.*, 1996, vol. A38, no. 2, pp. 210-215 (in Russian).
[135] Uryash, V.F., Grishatova, N.V., Kokurina, N.Yu., Gruzdeva, A.E., Faminskaya, L.A., Karyakin, N.V., Physical-chemical analysis of inulin-water mixtures, *Intern. confer, on*

physical-chemical analysis of liquid-phase systems: Proceedings, Saratov: SSUniv. Publ., 2003, pp. 160 (in Russian).

[136] Rabinovich, I.B., Uryash, V.F., and Mochalov, A.N., Method of determining solubility of low-molecular substances in polymers, *USSR Inventor's Certificate* no. 1603993, 1990. Priority of 07.07.88 (in Russian).

[137] Berg, L.G., Introduction into Thermography, Moscow: Nauka Publ., 1969 (in Russian).

[138] Uryash, V.F., Maslova, V.A., and Chizhikova, V.A., Thermal analysis of agar and agar-water mixtures as a polymeric carrier for drugs, *Biomater.-Liv. Syst. Inter.,* 1994, vol. 2, no. 2, pp. 71-77 (in Russian).

[139] Rabinovich, I.B., Khlyustova, T.B., Mochalov, A.N., and Kokurina, N.Yu., Calorimetric determination of solubility of ordinary and heavy water in wood cellulose, *Zhurn. Obshchei Khimii*, 1989, vol. 59, no. 6, pp. 1240-1244 (in Russian).

[140] Uryash, V.F., Gruzdeva, A.E., Kokurina, N.Yu., Grishatova, N.V., and Faminskaya, L.A., Calorimetric determination of water solubility in pectin and diagram of physical states of pectin-water system, *Vysokomolek. Soedin*., 2007, vol. A49, no. 9, pp. 1672-1678 (in Russian).

[141] Flyate, D.M. and Grunin, Yu.B., Investigation of water interaction with cellulose using nuclear magnetic relaxation technique, *Bumazhnaya Prom*., 1973, no. 10, pp. 1-3 (in Russian).

[142] Rogovin, Z.A., Chemistry of Cellulose, Moscow: Khimiya, 1972 (in Russian).

[143] Kruglitskii, N.N., Polishchuk, T.N., Privalko, V.P., Vyaz'mitina, O.M., Investigation of cellulose material interaction with water using the technique of differential scanning calorimetry, *Ukrainskii Khimich. Zhurn.,* 1985, vol. 511, no. 12, pp. 1250-1254 (in Ukrainian).

[144] Rabinovich, I.B., Isotopy Effect on Physical-Chemical Properties of Liquids, Moscow: Nauka, 1968 (in Russian).

[145] Papkov, S.P., Gel-Like State of Polymers, Moscow: Khimiya, 1974 (in Russian).

[146] Kozlov, P.V. and Papkov, S.P., Physical-Chemical Foundations of Polymer Plasticization, Moscow: Khimiya, 1982 (in Russian).

[147] Papkov, S.P., Phase and relaxation transition during molecular plasticization of amorphous-crystalline polymers, *Acta Polym.,* 1983, Bd. 34, H. 8, S. 477-481.

[148] Kozlov, P.V., Structural plasticization of polymers, *Acta Polym,*. 1983, Bd. 34, H. 8, S. 449-454.

[149] Chalykh, A.E., Gerasimova, V.K., and Mikhailova, Yu.M., Phase State Diagrams of Polymer Systems, Moscow: Yanus-K Publ., 1998 (in Russian).

[150] Karyakin, N.V., Principles of Chemical Thermodynamics: Textbook, Nizhny Novgorod: NNUniv. Press, 2003 (in Russian).

[151] Abdullaev, G.M.B., Zeinalov, B.K., Saryeva, S.A., Kerova, Kh.I., Ismailov, I.I., and Akhmedova, A.Kh., Technique of olive oil production, *RF Patent* no. 1211280, 1986, Priority of 19.08.83 (in Russian).

[152] Fedorov, P.N., Lukina, I.N., and Vedernikov, E.P., Technique of producing oil from presscake of quickbeam (*Sorbus aucuparia L.*), *RF Patent* no. 1761781, 1992, Priority of 19.12.90 (in Russian).

[153] Uryash, V.F. and Kirnus, L.M., Technique of producing oil extract of biologically active substances from fruit-berry raw, *RF Patent* no. 2148624, 2000, Priority of 17.12.93 (in Russian).

[154] Litvinyuk, N.Yu., Anisimova, K.V., and Anisimov, A.B., Technique of cryogenic freezing for subsequent sublimation drying in the flow of inert gas, *Storage and processing of agricultural raw*, 2008, no. 9, pp. 39–41 (in Russian).

[155] Gruzdeva, A.E., Potemkina, E.V., Grishatova, N.V., and Krot, A.R., Technique of producing food supplements from vegetable raw, *RF Patent* no. 2110194, 1998, Priority of 03.06.97 (in Russian).

[156] Gruzdeva, A.E., Potemkina, E.V., and Grishatova, N.V., Pelletized food supplement, *RF Patent* no. 2124300, 1999, Priority of 26.12.97 (in Russian).

[157] Gruzdeva, A.E., Potemkina, E.V., Pletneva, N.B., Grishatova, N.V., Dorofeichuk, V.G., and Kon', I.Ya., Food module composition, *RF Patent* no. 2124847, 1999, Priority of 26.12.97 (in Russian).

[158] Uryash, V.F., Kokurina, N.Yu., Grishatova, N.V., Gruzdeva, A.E., and Novoselova, N.V., Heat capacity of a set of dried vegetable products in the 293-323 K range, *Vestnik of Lobachevsky State University of Nizhni Novgorod*, Nizhny Novgorod: NNUniv. Press, 2007, no. 2, pp. 109-111.

[159] Nikolaev, P.N. and Rabinovich, I.B., Thermochemistry of isotope compounds. 1. Effect of hydrogen displacement with double-weight hydrogen on benzol heat capacity, *Tr. po khim. i khim. tekhnologii,* Gorki: GSUniv. Press, 1961, vyp. 2, pp. 242–250 (in Russian).

[160] Nikolaev, P.N., Rabinovich, I.B., Gal'perin, V.A., and Tsvetkov, V.G., Isotopic effect of heat capacity and compressibility of deuterocyclohexane, *Zhurn. Fiz. Khimii*, 1966, vol. 40, no. 5, pp. 1091-1097 (in Russian).

[161] Uryash, V.F., Maslova, V.A., Rabinovich, I.B., Molodovskaya, E.V., and Kanygina, E.L., Heat capacity and physical-chemical analysis of albumin and plasminogen, *Zhurn. Prikladnoi Khimii,* 1991, vol. 64, no. 7, pp. 1498–1503 (in Russian).

[162] Uryash, V.F., Maslova, V.A., and Kokurina, N.Yu., Low-temperature heat capacity of native and denaturated immunoglobulin, *Zhurn. Fiz. Khimii*, 1996, vol. 70, no. 10, pp. 1915–1918 (in Russian).

[163] Pichugin, A.A. and Tarasov, V.V., Supercritical extraction and perspectives of developing new internal-drainage processes, *Uspekhi Khimii*, 1991, vol. 60, no. 11, pp. 2412-2421 (in Russian).

[164] Kustov, L.M. and Beletskaya, I.P., "Green Chemistry" – new thinking, *Rossiiskii Khimicheskii Zhurnal (Zh. Ros. Khim. Ob-va im. D.I. Mendeleeva*), 2004, vol. 48, no. 6, pp. 3-12 (in Russian).

[165] Kas'yanov, G.I., Kriulin, V.P., and Leonchik, B.I., Methods and Technologies of CO_2-Extract Production, Moscow: AgroNIITEIPP Pudl., 1992 (in Russian).

[166] Gumerov, F.M., Sabirzyanov, A.N., and Gumerova, G.I., Sub- and Supercritical Fluids in Polymer processing, Kazan: FEN Publ., 2007 (in Russian).

[167] Latin, N.N. and Banashek, V.M., CO_2-extracts in product manufacturing, *Masla i Zhiry*, 2003, no. 1, pp. 6–7 (in Russian).

[168] Bogolitsyn, K.G., Modern trends in chemistry and chemical engineering of vegetable raw, *Rossiiskii Khimicheskii Zhurnal (Zh. Ros. Khim. Ob-va im. D. I. Mendeleeva*), 2004, vol. 48, no. 6, pp. 105-123 (in Russian).

[169] Sheldon, R.A., Catalytic transformations in water and critical carbon dioxide from the point of view of stable development conception, *Rossiiskii Khimicheskii Zhurnal (Zh. Ros. Khim. Ob-va im. D. I. Mendeleeva*), 2004, vol. 48, no. 6, pp. 74-83 (in Russian).

[170] The Role of Thermodynamics as a Basis of Development of Fundamental Investigations, Standartization and Technologies, Lebedev, Yu.A., Ed., Moscow: Akademkniga Publ., 2007 (in Russian).

[171] Kas'yanov, G.I., Liquid Carbon Dioxide as an Extragent of Aromatic and Biologically Active Substances of Vegetable Raw, Krasnodar: Krasnodarskoe Pravleniya NTO Pishchevoi Promyshlennosti Publ., 1980 (in Russian).

[172] Roslyakova, E.Yu. and Kalatskaya, L.M., CO_2-Extracts in Cosmetics, Krasnodar: KubGTUniv. Publ., 2001 (in Russian).

[173] Uryash, V.F., Gruzdeva, A.E., Kokurina, N.Yu., Grishatova, N.V., Uryash, A.V., and Karpova, I.G., Composition and physical-chemical properties of pine shoot extracts produced using supercritical fluid extraction, *Sverkhkriticheskie Flyuidy – Teoriya i Praktika,* 2008, vol. 3, no. 4 pp. 35-44 (in Russian).

[174] Gruzdeva, A.E., Grishatova, N.V., Demidik, M.M., Yakubova, I.Sh., Levachev, O.S., Chesnokova, T.A., Zakamennykh, T.N., Tyulina, N.E., and Solomakha, K.G., Methods of producing CO_2–extracts, *RF Patent* no. 2264442, 2005, Priority of 29.10.03 (in Russian).

[175] Markova, M.E., Uryash, V.F., Gruzdeva, A.E., Grishatova, N.V., Kokurina, N.Yu., and Uryash, A.V., Determination of water content in biological active additives for food prepared by supercritical fluid extraction. In. Biotechnology, Biodegradation, Water and Foodstuffs, Zaikov, G.E. and Krylova, L.P., Eds, NY: Nova Sci. Publ. Inc., 2009, pp. 101-106.

[176] Uryash, V.F., Rabinovich, I.B., Kutasova, G.A., and Ivanov, S.V., Method of determining water content in biopreparations, *RF Patent* no. 1814058, 1993, Priority of 27.08.90 (in Russian).

[177] Semenov, A.A., Natural Antitumor Compounds (Structure and Mechanism of Action), Gruntenko, E.V., Ed., Novosibirsk: Nauka Publ., 1979.

[178] Tolstikov, G.A., Flekhter, O.B., Shul'ts, E.E., Baltina, L.A., and Tolstikov, A.G., Betulin and its derivatives. Chemistry and biological activity, *Khimiya v Interesakh Ustoichivogo Razvitiya*, 2005, vol. 3, pp. 30 (in Russian).

[179] Tolstikova, T.G., Sorokina, I.V., Tolstikov, G.A., Tolstikov, A.G., and Flekhter, O.B., Lupan-group terpenoids – biological activity and pharmacological perspectives , *Bioorganicheskaya Khimiya*, 2006, vol. 32, no. 1, pp. 42–55 (in Russian).

[180] Flekhter, O.B., Medvedeva, N.I., Karachurina, L.T., Baltina, L.A., Galin, F.Z., Zarudii, F.S., and Tolstikov, G.A., Synthesis and pharmacological activity of betulin ethers, betulin acid and allobetulin, *Khimiko-Farmatsevticheskii Zhurnal*, 2005, vol. 39, no. 8, pp. 9–13 (in Russian).

[181] Uryash, V.F., Gruzdeva, A.E., Grishatova, N.V., Kokurina, N.Yu., Faminskaya, L.A., Larina, V.N., Uryash, A.V., and Kalashnikov, I.N., Physical-chemical properties of biologically active substances obtained using supercritical fluid extraction. *5-th International Conference "Supercritical fluids: fundamental basis, technologies, innovations": Proceedings,* Suzdal: Inst. Khimii Ra-ov RAN Publ., 2009, pp. 102 (in Russian).

[182] Uryash, V.F., Kokurina, N.Yu., Gruzdeva, A.E., Faminskaya, L.A., Uryash, A.V., and Kalashnikov, I.N., Physicochemical characteristics of betulin prepared by supercritical fluid extraction, *17-th International Conference on Chemical Thermodynamics in Russia: Proceedings,* Kazan: KSTUniv. Publ., 2009, vol. 1, pp. 472 (in Russian).

[183] Kuznetsov, B.N., Levdanskii, V.A., Es'kin, A.P., and Polezhaeva, N.I., Extraction of betulin and suberin from birch bark activated in conditions of "explosion autohydrolysis", *Khimiya Rastitel'nogo Syr'ya*, 1998, no. 1, pp. 5-9 (in Russian).

[184] Physical Chemistry: Principles and Applications in Biological Sciences, Tinoco, I., Sauer, K., Wang, J.C., Puglisi, J.D., Eds., NJ: Pearson Education, Inc. Publ., 2002.

[185] Boling, E.A., Blanchard, G.G., and Russel, W.J., Bacterial identification by microcalorimetry, *Nature*, 1973, vol. 241, no. 5390, pp. 472-473.

[186] Sikorenko, Yu.B., Rekharskii, M.V., Arens, A.K., Lopatnev, S.V., and Ozola, V.A., Enthalpy of enzymic hydrolysis of N-acetyl-D,L-methionine, *12-th All-Union conference on chemical thermodynamics and calorimetry: Proceedings,* Gorki: GSUniv. Press, 1988, Part.1, pp. 64 (in Russian).

[187] Gustafsson, L. and Larson, C., Energy budgeting in studying the effect of environmental factors on the energy metabolism of yeasts, *Thermochim. Acta.*, 1990, vol. 172, pp. 95-104.

[188] Liu Xiong, Yen Lung-Fei, Wang Baohuai, and Zhang Youmin, Microcalorimetric study on the interaction of F-actin with myosin and its proteolytic fragments, *Thermochim. Acta.*, 1995, vol. 253, pp. 167-174.

[189] Li Xie Chang, Hong Wang, and Sheng Qu Song, Microcalorimetric study on the aerobic growth of *Escherichia coli*, *Thermochim. Acta.*, 1995, vol. 253, pp. 175-182.

[190] Liu Yi, Feng Ying, Xie Changli, Qu Songsheng, Feng Changjian, Yue Zhifeng, Zhang Xiangcai, and Wu Zishen, Microcalorimetric investigation of the effect of D-glycosamine shiff base and its complexes on bacteria, *J. Wuhan Univ. Natur. Sci. Ed.*, 1995, vol. 41, no. 4, pp. 434-438 (in Chinese).

[191] Haynes, C.A. and Norde, W., Structures and stabilities of adsorbed proteins, *J. Colloid and Interface Sci.*, 1995, vol. 169, no. 2, pp. 313-328.

[192] Docolomansky, P., Breier, A., Gemeiner, P., and Ziegelhoffer, A., Screening of binding properties of con-A immobilized on bead cellulose by flow microcalorimetry using invertase and anti-con-A antibody as reporting systems, *Anal. Lett.*, 1995, vol. 28, no. 15, pp. 2585-2594.

[193] Yi Liang, Wang Cunxin, Wu Dingquan, and Qu Songsheng, Thermokinetic studies of the irreversible inhibition of single-substrate, enzyme-catalyzed reactions, *Thermochim. Acta.*, 1995, vol. 268, pp. 17-25.

[194] Kajiyama, K., Tomiyama, T., Uchiyama, S., and Kobayashi, Y., Phase transitions of sequenced polytripeptides observed by microcalorimetry, *Chem. Phys. Lett.*, 1995, vol. 247, no. 3, pp. 299-303.

[195] Liu Yi, Feng Ying, Xie Changli, Qu Songsheng, Feng Changjian, and Le Zhifeng, Thermochemical investigation of optimal bacteria growth, *Chin. J. Appl. Chem.*, 1996, vol. 13, no. 2, pp. 95-97 (in Chinese).

[196] Nan Zhaodong, Liu Yongjun, Sun Haitao, and Zhang Honglin, Microcalorimetric investigations of lowest bacteria growth temperatures, *Chin. J. Appl. Chem.*, 1996, vol. 13, no. 5, pp. 112-113 (in Chinese).

[197] Forte, L., Vinci, G., and Antonelli, M. L., Isothermal microcalorimetry as a useful tool for fat determination in food, *Anal. Lett.*, 1996, vol. 29, no. 13, pp. 2347-2362.

[198] Rowshan, H., Bordbar, A.K., and Moosavimovahedi, A.A., The thermodynamic stability of the different forms of β-Lactoglobulin (A and B) to sodium n-dodecyl sulfate, *Thermochim. Acta.*, 1996, vol. 285, no. 2, pp. 221-229.

[199] Bordbar, A.K., Moosavimovahedi, A.A., and Saboury, A.A., Comparative thermodynamic stability of bovin and pigeon hemoglobins by interaction with sodium N-dodecyl sulfate, *Thermochim. Acta.*, 1996, vol. 287, no. 2, pp. 343-349.

[200] Liang Yi, Wang Cun-Xin, Wu Ding-Quan, and Qu Song-Sheng, Application of microcalorimetry to product inhibition of single-substrate enzymic-catalytic reaction, *Acta Chim. Sin.*, 1996, vol. 54, no 1, pp. 38-44 (in Chinese).

[201] Feng Ying, Liu Yi, Xie Changli, Qu Songsheng, Le Zhifeng, Feng Changjian, Shen Haoyu, and Zhang Xangcai, Thermodynamic investigation of bacteria metabolism under the effect of shiff-base-type drugs, *J. Wuhan Univ. Natur. Sci. Ed.*, 1996, vol. 42, no. 4, pp. 423-428 (in Chinese).

[202] Bogomolov, A.A., Bikbov, T.M., Matveev, Yu.I., and Manakov, M.N., Conformational transformation of reserve proteins of soybean (*Glicine max*) in the process of seed germination, *Molekul. Biol.*, 1997, vol. 31, no. 1, pp. 91-97 (in Russian).

[203] Surin, A.K., Kotova, N.V., Marchenkova, S.Yu., Sokolovskii, I.V., Rodionova, N.A., Yaklichkin, S.Yu., and Semisotnov, G.V., Denaturation transitions of molecular schaperon GroEL from *Escherichia coli., Bioorgan. Khimiya*, 1997, vol. 23, no. 4, pp. 251-256 (in Russian).

[204] Czajkowska, D. and Witkowska-Gwiazdowska, A,. Postep w analizach mikrobiologicznych zywnosci. 2. Lastosowanie metod instrumentalnych, *Przem. Spoz.*, 1997, vol. 51, no. 4, pp. 39-42 (in Poland).

[205] Wang Tian-Zhi, Li Wei-Ping, Liy Yi, Wan Hong-Wen, Wu Ding-Quan, and Qu Song-Sheng, Investigation of direct catalytic enzyme-substrate reactions and analysis of transition states using microcalorimetry, *Acta Chim. Sin.*, 1998, vol. 56, no. 7, pp. 625-630 (in Chinese).

[206] Lamprecht, I., Monitoring metabolic activities of small animals by means of microcalorimetry, *Pure and Appl. Chem.*, 1998, vol. 70, no. 3, pp. 695-700.

[207] Yao Jun, Lin Yi, Tuo Yong, Liu Jianben, chen Xiong, Zhou Qin, Dong Jiaxin, Qu Songsheng, and Yu Zinin, The action of Cu^{2+} on *Bacillus thuringiensis* growth investigated by microcalorimetry, *Prikladnaya Biokhimiya i Mikrobiologiya*, 2003, vol. 39, no. 6, pp. 656-660 (in Russian).

[208] Ovchinnikov, Yu.A., Bioorganic Chemistry, Moscow: Prosveshchenie Publ., 1987 (in Russian).

[209] Uryash, V.F., Novoselova, N.V., Grishatova, N.V., and Gruzdeva, A.E., Thermodynamic investigation of polysaccharide enzymic hydrolysis, *14-th Intern. Conference on chemical thermodins: Proceedings,* St.-Petersburg: NIIKhimii SPbSUniv. Press, 2002, pp. 406 (in Russian).

[210] Grishatova, N.V., Uryash, V.F., Gruzdeva, A.E., and Novoselova, N.V., Calorimetric investigation of enthalpy of enzymic hydrolysis of vegetable raw product components for enteral feeding, *Zhurn. Prikladn. Khimii*, 2006, vol. 79, no. 12, pp. 1953-1957 (in Russian).

[211] Rakhmanov, R.S. and Gruzdeva, A.E., New balances natural concentrated food product for enteral tube feeding, *7-th Intern. Congress on parenteral and enteral feeding: Proceedings,* Moscow: 2003, pp. 81-82 (in Russian).

[212] Enzymes, Braunshtein, A.E., Ed., Moscow: Nauka, 1964 (in Russian).

[213] Bernhard, S.A., The Structure and Function of Enzymes, NY-Amsterdam: W.A. Benjamin. Inc. Publ., 1968.

[214] Kretovich, V.L., Introduction into Enzymology, Moscow: Nauka, 1986 (in Russian).
[215] Green, N.P.O., Stout, G.W., and Taylor, D.J., Biological Science, Soper, R., Ed., Cambridge Eng.: Cambridge Univ. Press, 1984, vol. 1.
[216] Musil, J., Novakova, O., and Kunz, K., Biochemistry in Schematic Perspective, Prague: Avicenum Czech. Med. Press, 1980 (in Czechoslovak).
[217] Radbil', O.S., Pharmacological Therapy in Gastroenterology: Reference book, Moscow: Meditsina, 1991 (in Russian).
[218] Methods of Biochemical Investigation of Plants, Ermakova, A.I., Ed., Leningrad: Agropromizdat Publ., 1987 (in Russian).
[219] Gal'perin, L.N., Kolesov, Yu.R., Mashkinov, L.B., and Germer, Yu.E., Multipurpose differential automated calorimeters (DAC), *6-th All-Union conference on calorimetry: Proceedings,* Tbilisi: Metsniereba Publ., 1973, pp. 539-543 (in Georgian).
[220] Gracheva, I.M. and Krivova, A.Yu., Technology of Enzymic Preparations, Moscow: Elevar Publ., 2000 (in Russian).
[221] Andreeva, Z.M., Bendas, L.G., El'chinova, E.A., and Shepilova, R.G., Methodological Recommendations to Control Biological Parameters of Nutrient Media, Moscow: Minzdrav SSSR Publ., 1980 (in Russian).
[222] Development and Implementation of Scientifically Grounded Recommendations on Standardizationof Nutrient Media Production Technologies, N.O. Emelyanovs, Ed., *RandD Report, State Registration no. 01.86.0031254,* Shchelkovo: Ros. Nauchno-Tekhnich. Inform. Tsentr Publ., 1989 (in Russian).
[223] Green, N.P.O., Stout, G.W., and Taylor, D.J., Biological Science, Soper, R., Ed., Cambridge Eng.: Cambridge Univ. Press, 1985, vol. 3.
[224] Uryash, V.F., Gorlova, I.S., Novoselova, N.V., and Kon'kova, N.K., Thermochemical investigation of lactobacilli cultivation on different nutrient media, *Zhurn. Fiz. Khimii*, 2010, vol. 84 (in press) (in Russian).
[225] Ivanov, V.N., Energy of Microorganism Growth (Investigation of Microorganism Vital Functions Based on the Balance of Microenergetic Compounds), Kiev: Nauk. Dumka Publ., 1981 (in Ukrainian).
[226] Boling, E.A., Blanchard, G.G., and Russel, W.J., Bacterial identification by microcalorimetry, *Nature*, 1973, vol. 241, no. 5390, pp. 472-473.
[227] Gustafsson, L., Microbiological calorimetry, *Thermochim. Acta.*, 1991, vol. 193, pp. 145-171.
[228] Popov, G.A., Polivoda, B.I., and Konev, V.V., Temperature phase transitions in neoplastic cells, *Biofizika*, 1980, vol. 25, no. 1, pp. 177-178 (in Russian).
[229] Andronikashvili, E.L., Mrevlishvili, G.M., Bakradze, N.G., Madzhagaladze, G.V., Monaselidze, D.R., Chanchalashvili, Z.I., Calorimetric investigation of the nature of intermolecular melting of collagen extracted from transplantable tumor, *Dokl. AN SSSR*, 1968, vol. 183, no. 1, pp. 212-214 (in Russian).
[230] Andronikashvili, E.L. and Mrevlishvili, G.M., Investigation of the state of water in tumor tissues using calorimetric technique, *Dokl. AN SSSR*, 1968, vol. 183, no. 2, pp. 463-465 (in Russian).
[231] Andronikashvili, E.L., Monaselidze, D.R., Chanchalashvili, Z.I., and Madzhagaladze, G.V., Cooperative thermal transitions in normal and neoplastic cells, *Biofizika*, 1983, vol. 28, no. 3, pp. 528-537 (in Russian).

[232] Bihari – Varga M., The application of thermoanalytical methods in studies on the pathomechanism of human diseases, *J. Therm. Analys.*, 1992, vol. 38, pp. 153-157.
[233] Herrmann, A., Arnold, K., Lassmann, G. Glaser R., Structural transitions of the erythrocyte membrane: An ESR approach, *Acta Biol. Med. Germ.,* 1982, Bd. 41, H. 4, S. 289-298.
[234] Kalashnikov, S.P., Uryash, V.F., Fokin, V.M., and Uryash, A.V., Utilization of DTA for studying erythrocytes and their component in the −190 – +230°C range, *Nizhegorodski meditsinskii zhurnal*, Nizhny Novgorod: NSMedAc. Press, 1996, no. 3, pp. 5-8 (in Russian).
[235] Biochemical Analysis of Membranes, Maddy, A.H., Ed., London: Chapman and Hall Ltd. Press, 1976.
[236] Uryash, V.F., Maslova, V.A., and Kon'kova, N.K., Physicochemical analysis of gelatin-water system, *Biomater.-Liv. Syst. Inter.*, 1995, vol. 3, nos. 1-2, pp. 39-44.
[237] Perova, N.V., Porunova, Yu.V., Uryash, V.F., Faminskaya, L.A., Krasheninnikov, M.E., Rasulov, M.F., Onishchenko, N.A., Sevast'yanov, V.I., and Shumakov, V.I., Biodegraded collagen-containing matrix Spherogel™ for bioartificial organs and tissues, *Perspectivnye Materialy*, 2004, no. 2, pp. 52-59 (in Russian).
[238] Uryash, V.F., Sevast'yanov, V.I., Kokurina, N.Yu., Porunova, Yu.A., and Faminskaya, L.A., Heat capacity, physical transitions of collagen and solubility of water in it, *Zhurn. Obshchei Khimii*, 2006, vol. 76, no. 9, pp. 1421-1425 (in Russian).
[239] Liquid Crystals, Zhdanov, S.I., Ed., Moscow: Khimiya, 1979 (in Russian).
[240] Mints, R.I. and Kononenko, E.V., Liquid crystals in biological systems, *Itogi nauki I tekhniki. Biofizika*, Moscow: Vses. Inst. Nauchno-Tekhnich. Inform. Publ., 1982, vol. 13, pp, 3-150 (in Russian).
[241] Dowson, R.M.C., Elliot, D.C., Elliot, W.H., and Jones, K.M., Data for Biochemical Research, Oxford: Clarendon Press, 1986.
[242] Yagfarov, M.Sh., New method of measuring heat capacities and thermal effects, *Zhurn. Fiz. Khimii,* 1968, vol. 43, no. 6, pp. 1620-1625 (in Russian).
[243] Kabo, F.G. and Diky, V.V., Details of calibration of a scanning calorimeter of the triple heat bridge type, *Thermochim. Acta.*, 2000, vol. 347, pp. 79-84.
[244] Uryash, A.V., Bulanov, G.A., Gordetsov, A.S., Tsybusov, S.N., and Uryash, V.F., Method of diagnosing malignant neoplasts, *RF Patent* no. 2140638, 1999, Priority of 19.03.98 (in Russian).
[245] Gorskii, S.M. and Il'icheva, K.V., Spectrofluorimetry and spectrophotometry of biological fluids, *4-th All-Union workshop on application of luminescence analysis in medicine and biology and its hardware implementation: Proceedings,* Moscow: 1992, p. 9 (in Russian).
[246] Kukosh, V.I., Gordetsov, A.S., Mushkin, Yu.I., Skobeleva, S.E., Pavlova, E.K., Uchugina, A.F., Mamaev, Yu.P., Latyaeva, V.N., and Dergunov, Yu.I., Methods of lung carcinoma diagnostics, *RF Patent* no. 1489373, 1995, Priority of 19.05.87 (in Russian).
[247] Katsnel'son, B.A., Degtyareva, T.D., and Privalova, L.I., Biological prophylactics of intoxications with nonorganic substances, *Meditsina Truda i Promyshlennaya Ekologiya*, 2004, no. 9, pp. 19–23 (in Russian).

[248] Degtyareva, T.D., Katsnel'son, B.A., and Privalova, L.I., Evaluation of efficiency of biological means of lead intoxication prophylactics (experimental investigation), *Meditsina Truda,* 2000, no. 3, pp. 40–43 (in Russian).

[249] Dudkin, M.S. and Shchelkunov, L.F., Dietary fibers and new food products, *Voprosy Pitaniya*, 1998, vol. 67, no. 1, pp. 35-38 (in Russian).

[250] Dudkin, M.S. and Shchelkunov, L.F., Dietary fibers – a new section of food chemistry and technology, *Voprosy Pitaniya*, 1998, vol. 67, no. 2, pp. 36-38 (in Russian).

[251] Alternative Medicine. Non-Medicamentous Methods of Treatment, Belyakova, N.A., Ed., Arkhangel'sk: Sev.-Zap. Knizhn. Press, 1994 (in Russian).

[252] Gaev, P.A., Kalev, O.F., and Korobkin, A.V., Enterosorption as a Method of Efferent Therapy, Chelyabinsk: ChelyabSMedAcad. Publ., 2001 (in Russian).

[253] Shchelkunov, L.F., Dudkin, M.S., and Danilova, E.I., Dietary fibers as enterosorbents of ecologically hazardous substances, *2-nd Intern. Seminar "Ecology of a man: problems and state of clinical and prophylactic nutrition": Proceedings,* Pyatigorsk: Pyatigorskaya St. Pharm. Acad. Press, 1993, pp. 31–32 (in Russian).

[254] Stavitskaya, S.S., Mironyuk, T.I., Kartel', N.T., and Strelko, V.V., Sorption properties of "dietary fibers" in secondary products of vegetable raw processing, *Zhurn. Prikladnoi Khimii*, 2001, vol. 74, no. 4, pp. 575–578 (in Russian).

[255] Pogozhaeva, A.V., Dietary fibers in clinical-prophylactic nutrition, *Lechebnoe Pitanie*, 1998, no. 3, pp. 20–22 (in Russian).

[256] Vainshtein, S.G. and Masik, A.M., Dietary fibers and nutrient, *Voprosy Pitaniya*, 1984, vol. 53, no. 3, pp. 6–12 (in Russian).

[257] Beriketov, A.S., Oitov, Z.Kh., and Guchinov, V.A., Methodological recommendations on using pectin and pectin-containing products for treatment and prophylactics, Pyatigorsk: Pyatigorskaya St. Pharm. Acad. Press, 1999 (in Russian).

[258] Zolotareva, A.M., Chirkina, T.F., and Tsibikova, D.Ts., Investigation of functional properties of sea-buckthorn pectin, *Khimiya Rastitel'nogo Syr'ya*, 1998, no. 1, pp. 29–21 (in Russian).

[259] Trakhtenberg, I.M., Krasnyuk, E.P., and Lubyanova, I.P., Mechanism of Action and Pharmacology of Pectin. Clinical Approbation of "Medetopect" as a Means of Individual Prophylactics in Case of Professional Exposure to Lead Action, Kiev: Institut Meditsiny Truda AMN Ukrainy Press, 1996 (in Ukrainian).

[260] Zelenin, K.N., Chelates in medicine, *Sorosovskii Obrazovatel'nyi Zhurnal*, 2001, vol. 7, no. 1, pp. 45–50 (in Russian).

[261] Benguella, B. and Benaissa, H., Cadmium removal from aqueous solutions by chitin: kinetic and equilibrium studies, *Water Research*, 2004, vol. 36, no. 10, pp. 2463–2474.

[262] Uryash, V.F., Gruzdeva, A.E., Pletneva, N.B., Maslova, E.A., Potemkina, E.V., Demarin, V.T., Investigation of lead and cadmium sorption with a number of vegetable raw products, *Khimiya, tekhnologiya i promyshlennaya ekologiya neorganicheskikh soedinenii*, Perm: Perm. SUniv. Press, 1999, vyp. 2, pp. 56-59 (in Russian).

[263] Uryash, V.F., Stepanova, E.A., Grishatova, N.V., Gruzdeva, A.E., Kuleshova, N.V., and Bezrukov, M.E., Investigation of heavy metal sorption by "Biofit" food supplements, *Vestnik of Lobachevsky State University of Nizhni Novgorod. Seriya Biologiya*, Nizhny Novgorod. NNUniv. Press, 2004, no. 3(5), pp. 85-91 (in Russian).

[264] Stepanov, E.A., Uryash, V.F., Silkin, A.A., Loginov, V.V., Gruzdeva, A.E., Grishatova, N.V., and Tumanova, A.N., Investigation of lead sorption and excretion by biologically

active food supplements in *in vitro* and *in vivo* experiments, *Povolzhskii Ekologicheskii Zhurnal,* 2005, no. 1, pp. 71-75 (in Russian).

[265] Stepanov, E.A., Sorption of lead and cadmium by biologically active food supplements from vegetable raw in bioprophylactics of environmental pollution with heavy metals Candidate (Biology) Dissertation (PhD), Nizhny Novgorod: NNUniv., 2006 (in Russian).

[266] Markova, M.E., Stepanov, E.A., Uryash, V.F., Gruzdeva, A.E., Grishatova, N.V., Demarin, V.T., and Tumanova, A.N., Sorption of heavy metals with higher fungi and chitosan of different origin in in vitro experiments. *Vestnik of Lobachevsky State University of Nizhni Novgorod,* Nizhny Novgorod: NNUniv. Press, 2008, no. 6, pp. 118-124 (in Russian).

[267] Kosyakov, V.N., Veleshko, I.E., Yakovlev, N.G., Rozanov, K.V., and Gorovoi, L.F., Sorption of radioactive nuclide by chitin sorbents of different origin, *7-th Intern. Confer. "Modern trends in chitin and chitisan investigations": Proceedings,* St.-Petersburg: VNII Rybnogo Khoz. i Okeanografii Publ., 2003, pp. 320-323 (in Russian).

[268] Solodovnik, T.V., Unrod, V.I., Minaev, B.F., and Pakhar', S.A., Theoretical investigation of complex formation mechanism in the chitin-Pb(II) system, *8-th Intern. Conference "Modern trends in chitin and chitisan investigations": Proceedings,* Moscow: VNII Rybnogo Khoz. i Okeanografii Publ., 2006, pp. 130-132 (in Russian).

[269] Voronova, M.I. and Zakharov, A.G., Adsorption of phenol and toluene from the gas phase and water solutions on cellulose, *Zhurn. Prikladnoi Khimii,* 2009, vol. 82, no. 3, pp. 410-413 (in Russian).

[270] Kosobutskaya, A.A., Naimark, N.I., and Tarakanov, O.G., Sorption of gaseous ammonia by cellulose acetates in a wide range of displacement degrees, *Vysokomolek. Soedin.,* 1983, vol. B5, no. 1, pp. 18-22 (in Russian).

[271] Mudarisova, R.Kh., Kulish, E.I., Zinatullin, R.M., Tamindarova, N.E., Kolesov, S.V., Khunafin, S.N., and Monakov, Yu.B., Films of chitosan-based complexes with controlled levomecetine release, *Zhurn. Prikladnoi Khimii,* 2006, vol. 79, no. 10, pp. 1737-1739 (in Russian).

[272] Nikiforova, T.E., Kozlov, V.A., Bagrovskaya, N.A., and Rodionova, M.V., Sorption properties of enzymically modified flax fiber, *Zhurn. Prikladnoi Khimii,* 2007, vol. 80, no. 2, pp. 236-241 (in Russian).

[273] ASTM E2456 - 06 Standard Terminology Relating to Nanotechnology, *Book of Standards,* 2006.

[274] Bogunia-Kubik, K., and Sugisaka M., From molecular biology to nanotechnology and nanomedicine, *Bio Systems,* 2002, vol. 65, pp. 123-138.

[275] Buxton, D., The promise of nanotechnology for heart, lung and blood diseases. *Expert Opin Drug Deliv.,* 2006, vol. 3, pp. 173-175.

[276] Buxton, D.B., Lee, S.C., Wickline, S.A., and Ferrari, M., Recommendations of the National Heart, Lung, and Blood Institute Nanotechnology Working Group, *Circulation,* 2003, vol. 108, pp. 2737-2742.

[277] Fernandez, P.L., Nanotechnology, nanomedicine and nanopharmacology. *An. R. Acad. Nac. Med. (Madr.),* 2007, vol. 124, pp. 189-200.

[278] Kameneva, M.V., Wu, Z.J., Uraysh, A., Repko, B., Litwak, K.N., Billiar, T.R., Fink, M.P., Simmons, R.L., Griffith, B.P., and Borovetz, H.S., Blood soluble drag-reducing

polymers prevent lethality from hemorrhagic shock in acute animal experiments, *Biorheology,* 2004, vol. 41, pp. 53-64.

[279] Antonova, N., and Lazarov, Z., Hemorheological and hemodynamic effects of high molecular weight polyethylene oxide solutions, *Clinical Hemorheology a. Microcirculation,* 2004, vol. 30, pp. 381-390.

[280] Nishiyama, N., and Kataoka, K., Current state, achievements, and future prospects of polymeric micelles as nanocarriers for drug and gene delivery, *Pharmacol. Ther.,* 2006, vol. 112, pp. 630-648.

[281] Nishiyama, N., and Kataoka, K., Medical applications of nanotechnology: polymeric micelles for drug delivery, *Nippon Geka Gakkai Zasshi,* 2005, vol. 106, pp. 700-705.

[282] Pacella, J.J., Kameneva, M.V., Csikari, M., Lu, E., and Villanueva, F.S., A novel hydrodynamic approach to the treatment of coronary artery disease, *European Heart J.,* 2006, vol. 27, pp. 2362-2369.

[283] Pierstorff, E., and Ho, D., Monitoring, diagnostic, and therapeutic technologies for nanoscale medicine, *J. Nanoscience a. Nanotech.,* 2007, vol. 7, pp. 2949-2968.

[284] Salata, O., Applications of nanoparticles in biology and medicine, *J. Nanobiotech.,* 2004, vol. 2, pp. 3-7.

[285] Scott, N.R., Nanotechnology and animal health. *Rev. Sci. Tech. (International Office of Epizootics),* 2005, vol. 24, pp. 425-432.

[286] Suri, S.S., Fenniri, H., and Singh, B., Nanotechnology-based drug delivery systems. *J. Occup Med. Toxicol.,* 2007, vol. 2, pp. 16-20.

[287] Aliabadi, H.M., Elhasi, S., Brocks, D.R., and Lavasanifar, A., Polymeric micellar delivery reduces kidney distribution and nephrotoxic effects of cyclosporine A after multiple dosing, *J. Pharm. Sci.,* 2008, vol. 97, pp. 1916-1926.

[288] Cotoia, A.K.M., Marascalco, P.J., Fink, M.P., Delude, R.L., Drag-reducing hyaluronic acid increases survival in profoundly hemorrhaged rast, *Shock,* 2009, vol. 31, pp. 258-261.

[289] De Campos, A.M., Sanchez, A., Gref, R., Calvo, P., and Alonso, M.J., The effect of a PEG versus a chitosan coating on the interaction of drug colloidal carriers with the ocular mucosa, *Eur. J. Pharm. Sci.,* 2003, vol. 20, pp. 73-81.

[290] Dziubla, T.D., Shuvaev, V.V., Hong, N.K., Hawkins, B.J., Madesh, M., Takano, H., Simone, E., Nakada, M.T., Fisher, A., Albelda, S.M., and Muzykantov, V.R., Endothelial targeting of semi-permeable polymer nanocarriers for enzyme therapies, *Biomaterials,* 2008, vol. 29, pp.215-227.

[291] Gutierrez, G., and Fuller, S.P., Of hemorrhagic shock, spherical cows and *Aloe Vera. Critical Care (London, England),* 2004, vol. 8, pp. 406-407.

[292] Han, H.D., Lee, A., Hwang, T., Song, C.K., Seong, H., Hyun, J., and Shin, B.C., Enhanced circulation time and antitumor activity of doxorubicin by comblike polymer-incorporated liposomes, *J, Control Release,* 2007, vol. 120, pp. 161-168.

[293] Klossner, R.R., Queen, H.A., Coughlin, A.J., and Krause, W.E., Correlation of chitosan's rheological properties and its ability to electrospin, *Biomacromolecules,* 2008, vol. 9, pp. 2947-2953.

[294] Liu, J., Zeng, F., and Allen, C., In vivo fate of unimers and micelles of a poly(ethylene glycol)-block-poly(caprolactone) copolymer in mice following intravenous administration, *Eur. J. Pharm. Biopharm.,* 2007, vol. 65, pp. 309-319.

[295] Macias, C.A., Kameneva, M.V., Tenhunen, J.J., Puyana, J.C., and Fink, M.P., Survival in a rat model of lethal hemorrhagic shock is prolonged following resuscitation with a small volume of a solution containing a drag-reducing polymer derived from *Aloe Vera, Shock,* 2004, vol. 22, pp. 151-156.

[296] Marhefka, J.N., Marascalco, P.J., Chapman, T.M., Russell, A.J., and Kameneva, M.V., Poly(N-vinylformamide) a drag-reducing polymer for biomedical applications, *Biomacromolecules,* 2006, vol. 7, pp. 1597-1603.

[297] McCloskey, C.A., Kameneva, M.V., Uryash, A., Gallo, D.J., and Billiar, T.R., Tissue hypoxia activates JNK in the liver during hemorrhagic shock, *Shock,* 2004, vol. 22, pp. 380-386.

[298] Polimeni, P.I., Ottenbreit, B., and Coleman, P., Enhancement of aortic blood flow with a linear anionic macropolymer of extraordinary molecular length, *J. Molecular a. Cellular Cardiology,* 1985, vol. 17, pp. 721-724.

[299] Sakai, T., Repko, B.M., Griffith, B.P., Waters, J.H., and Kameneva, M.V., I.V. infusion of a drag-reducing polymer extracted from aloe vera prolonged survival time in a rat model of acute myocardial ischaemia, *British J. Anaesthesia,* 2007, vol. 98, pp. 23-28.

[300] Sawchuk, A.P., Unthank, J.L., and Dalsing, M.C., Drag reducing polymers may decrease atherosclerosis by increasing shear in areas normally exposed to low shear stress, *J. Vasc. Surg.,* 1999, vol. 30, pp. 761-764.

[301] Wan, L., Zhang, X., Gunaseelan, S., Pooyan, S., Debrah, O., Leibowitz, M.J., Rabson, A.B., Stein, S., and Sinko, P.J., Novel multi-component nanopharmaceuticals derived from poly(ethylene) glycol, retro-inverso-tat nonapeptide and saquinavir demonstrate combined anti-HIV effects. *AIDS Res. Ther.,* 2006, vol. 3, pp.12-18.

[302] ZJ Wu, K.S., Marascalco, P., Marhefka, J., Kameneva, M.V., Modification of flow behavior of red blood cells by blood soluble drag-reducing polymers, *International Congress on Biological and Medical Engineering: Proceedings,* Singapore: 2002. p. 152.

[303] Bene, L., Szentesi, G., Matyus, L., Gaspar, R., and Damjanovich, S., Nanoparticle energy transfer on the cell surface, *J. Mol. Recognit.,* 2005, vpl. 18, pp. 236-253.

[304] Cohen, H., Levy, R.J., Gao, J., Fishbein, I., Kousaev, V., Sosnowski, S., Slomkowski, S., and Golomb, G., Sustained delivery and expression of DNA encapsulated in polymeric nanoparticles, *Gene Ther.,* 2000, vol. 7, pp. 1896-1905.

[305] Dalby, M.J., Marshall, G.E., Johnstone, H.J., Affrossman, S., and Riehle, M.O., Interactions of human blood and tissue cell types with 95-nm-high nanotopography, *IEEE Trans. Nanobioscience,* 2002, vol.1, pp. 18-23.

[306] Fernandes, J.C., Eaton, P., Nascimento, H., Belo, L., Rocha, S., Vitorino, R., Amado, F., Gomes, J., Santos-Silva, A., Pintado, M.E., and Malcata, F.X., Effects of chitooligosaccharides on human red blood cell morphology and membrane protein structure, *Biomacromolecules*, 2008, vol. 9, no. 12, pp.3346-3352.

[307] Greene, H.L., Nokes, R.F., and Thomas, L.C., Biomedical implications of drag reducing agents, *Biorheology,* 1971, vol. 7, pp. 221-223.

[308] Kim, D.H., Kim, K.N., Kim, K.M., and Lee, Y.K., Targeting to carcinoma cells with chitosan- and starch-coated magnetic nanoparticles for magnetic hyperthermia, *J. Biomed. Mater. Res.,* ser. A, 2009, vol. 88, pp. 1-11.

[309] Lim S.H., Liao, I.C., and Leong, K.W., Nonviral gene delivery from nonwoven fibrous scaffolds fabricated by interfacial complexation of polyelectrolytes, *Mol. Ther.,* 2006, vol. 13, pp. 1163-1172.

[310] Marconi, A., Fine, ultrafine and nano-particles in the living and working setting: potential health effects and measurement of inhalation exposure. *Giornale italiano di medicina del lavoro ed ergonomia,* 2006, vol. 28, pp. 258-265.

[311] Masuoka, K., Ishihara, M., Asazuma, T., Hattori, H., Matsui, T., Takase, B., Kanatani, Y., Fujita, M., Saito, Y., Yura, H., Fujikawa, K., and Nemoto, K., The interaction of chitosan with fibroblast growth factor-2 and its protection from inactivation, *Biomaterials,* 2005, vol. 26, pp. 3277-3284.

[312] Minelli, C., Kikuta, A., Tsud, N., Ball, M.D., and Yamamoto, A., A micro-fluidic study of whole blood behaviour on PMMA topographical nanostructures, *J. Nanobiotechnol.,* 2008, vol. 6, pp. 3-8.

[313] Mostardi, R.A., Greene, H.L., Nokes, R.F., Thomas, L.C., and Lue, T., The effect of drag reducing agents on stenotic flow disturbances in dogs, *Biorheology,* 1976, vol. 13, pp. 137-141.

[314] Mostardi, R.A., Thomas, L.C., Greene, H.L., VanEssen, F., and Nokes, R.F., Suppression of atherosclerosis in rabbits using drag reducing polymers, *Biorheology,* 1978, vol. 15, pp. 1-14.

[315] Muro, S., Garnacho, C., Champion, J.A., Leferovich, J., Gajewski, C., Schuchman, E.H., Mitragotri, S., and Muzykantov, V.R., Control of endothelial targeting and intracellular delivery of therapeutic enzymes by modulating the size and shape of ICAM-1-targeted carriers. *Mol. Ther.,* 2008, vol. 16, pp. 1450-1458.

[316] Wang, H., Hu, Y., Sun, W., and Xie, C., Polylactic acid nanoparticles targeted to brain microvascular endothelial cells. *J. Huazhong Univ. Sci. Technol. Med. Sci.,* 2005, vol. 25, pp. 642-644.

[317] Handy, R.D., Owen, R., and Valsami-Jones, E., The ecotoxicology of nanoparticles and nanomaterials: current status, knowledge gaps, challenges, and future needs, *Ecotoxicology (London, England),* 2008, vol. 17, pp. 315-325.

[318] Morgan, K., Development of a preliminary framework for informing the risk analysis and risk management of nanoparticles, *Risk. Anal.,* 2005, vol. 25, pp. 1621-1635.

[319] Oesterling, E., Chopra, N., Gavalas, V., Arzuaga, X., Lim, E.J., Sultana, R., Butterfield, D.A., Bachas, L., and Hennig, B., Alumina nanoparticles induce expression of endothelial cell adhesion molecules, *Toxicol. Lett.,* 2008, vol. 178, pp. 160-166.

[320] Singh, S. and Nalwa, H.S., Nanotechnology and health safety-toxicity and risk assessments of nanostructured materials on human health, *J. Nanoscience a. Nanotechnol.,* 2007, vol. 7, pp. 3048-3070.

[321] Zhang, C., Qu, G., Sun, Y., Yang, T., Yao, Z., Shen, W., Shen, Z., Ding, Q., Zhou, H., and Ping, Q., Biological evaluation of N-octyl-O-sulfate chitosan as a new nano-carrier of intravenous drugs, *Eur. J. Pharm. Sci.,* 2008, vol. 33, pp. 415-423.

[322] Polimeni, P.I., and Ottenbreit, B.T., Hemodynamic effects of a poly(ethylene oxide) drag-reducing polymer, Polyox WSR N-60K, in the open-chest rat, *J. Cardiovascular Pharm.,* 1989, vol. 14, pp. 374-380.

[323] Schumer, W., and Lefer, A., Molecular a. Cellular Aspects Shock a. *Trauma,* NY: Allan R. Liss, Inc., 1983.

[324] Gillissen, J.J., Polymer flexibility and turbulent drag reduction, *Phys. Rev. E.* 2008, vol. 78, pp. 046311-046318.
[325] Marhefka, J.N., Chapman, T.M., Kameneva, M.V., Mechanical degradation of drag reducing polymers in suspensions of blood cells and rigid particles, *Biorheology,* 2008, vol. 45, pp. 599-609.

Reviewed by
Svetlana Bulgakova, DrSci. Chem,
Chief Laboratory of Polymerization,
Research Institute of Chemistry Nizhni Novgorod State University.

In: Polymer Research and Applications
Editors: Andrew J. Fusco and Henry W. Lewis
ISBN: 978-1-61209-029-0

Chapter 8

ORGANOMETALLIC POLYAMINES AS PHYSICAL MATERIALS AND BIOMATERIALS

Charles E. Carraher, Jr. * ***and Amitabh Battin***
Florida Atlantic University, Department of Chemistry
and Biochemistry, Boca Raton, FL, USA
and Florida Center for Environmental Studies,
Palm Beach Gardens, FL, USA

ABSTRACT

A number of metal-containing polyamines have been synthesized employing the condensation process, generally one of the interfacial polycondensation processes. The majority of these have been synthesized from the reaction between mono and diamines and Group IVB metallocene dihalides and organotin dihalides. The major reasons for the synthesis of the Group IVB polyamines is to control light and as control release agents and more recently for electrical applications. Ruthenium-containing polyamines were synthesized as part of a solar energy conversion effort. The major reasons for the synthesis of the organotin polyamines is biological since many of these materials offer good ability to resist a wide variety of bacteria, cancer cell lines, and viruses.

Keywords: Metallocenes, ruthenium-containing polymers, metallocene-containing polymers, solar energy conversion, control-release, light control, organotin, arsenic-containing polymers, organosilicon-containing polymers, tin-containing polymer, pancreatic cancer, herpes.

INTRODUCTION

Metal-containing polymers are synthesized not only to create new materials but also to take advantage of the new, often essential, properties introduced because of the presence of the

* E-mail address: carraher@fau.edu

metal-containing moiety. This is the case with the metal-containing polyamines described in this paper.

The polyamines described here were generally synthesized employing one of several interfacial polycondensation processes. The interfacial process was popularized by Morgan and Carraher in the 1960s and 1970s and is today employed in the industrial synthesis of polycarbonates and aromatic polyamides generally known as aramids [1-4]. This process requires what is referred to as high energy reactants, namely reactants that allow reaction to occur at somewhat low activation energies, generally in the range of 10-20 Kcal/mole (40 to 80 KJ/mole). For polyamine synthesis, the amine-containing reactant is dissolved in one solvent, typically water, along with an added base such as sodium hydroxide. The metal-containing reactant is dissolved in a largely water immiscible liquid such as hexane. The two phases are brought together with rapid stirring and the polymer is typically formed within less than a minute.

The metal-containing polyamines are named with the metal-containing moiety treated as a methylene unit.

Here, the paper is divided into two major sections related to the particular metal that is present in the polyamine. The initial section deals with polyamines containing transition metals, generally Group IVB metallocenes. The second section describes polyamines containing main group metals, generally Group IVA and VA metals.

Transition Metal Metallocenes

General

Interest in metallocene compounds began in the 1950s when two independent groups prepared ferrocene by reaction of iron II chloride with cyclopentadienyl magnesium bromide and by reaction of reduced iron with cyclopentadiene in the presence of potassium oxide.[5,6] Ferrocene is a true sandwich compound with the iron resting between two flat-plained cyclopendadiene groups [7-12].

By comparison, the Groups IV B and V B metallocenes have the cyclopentadiene groups facing the metal atom but present in a distorted tetrahedral arrangement as **1**.[7]

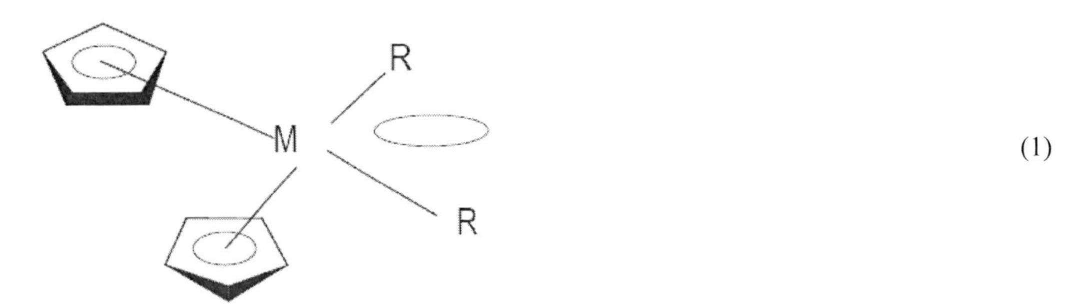

Cotton and Wilkinson [7] describe Group IVB metallocene compounds as containing 9-coordinate bonding with the hybrid orbitals being derived from one-s, three-p and five-d orbitals. Each pi-Cp ring involves three hybride orbitals. The remaining three orbitals consist of two equivalent $spd_{x^2-y^2}$, d_{z^2} orbitals (overlapping the halides, oxygen, nitrogen, etc.) and one sp vacant orbital as shown above, **1**.

Over 2 million entries are given when entering the terms hafnocene and polymers in SciFinder Scholar.[13] Almost all entries for Group IVB and VB metallocenes connected with polymers are associated with their use as stereoregulating catalysts. These metallocenes are very important Ziegler-Natta catalysts and are also important catalysts in the current generation of so-called soluble catalysts.[14]

While the amount of interest in these metallocenes is great, the amount of work involving their incorporation in polymers is relatively small with much of it involving metallocenes being present as part of a supported catalysis system. This may be due to the impression that the mechanism of catalytic activity involves having at least one unoccupied orbital available about the metal atom. For instance, Grubbs and co-workers in 1977 suggested, based on work done with olefin hydrogenations, that the two chlorides in titanocene dichloride are absent in the active catalyst.[15,16]

Halides are more electronegative than the metallocene Cp_2M portion so they can undergo condensation reactions where the metallocene dihalide acts as a Lewis acid. When the Lewis base is mono-functional the products will be monomeric as shown below for a metallocene dichloride reactions with a primary amine.

$$Cp_2MCl_2 + 2H_2NR \longrightarrow Cp_2M(RNH)_2 + 2HCl$$ Equation 1.

But when the Lewis base, here a diamine, is difunctional a chain is formed.

$$Cp_2MCl_2 + H_2NRNH_2 \longrightarrow \text{-(-(}Cp_2\text{)MHN-R-NH-)-} + 2HCl$$ Equation 2.

Group IVB Metallocene Polyamines

A number of titanocene polyamines have been synthesized including those derived from aromatic and aliphatic diamines, **2**.[17-20] Polyamines were synthesized using both the regular interfacial synthetic process and the aqueous solution process where both the diamines and metallocene dichloride are simply dissolved in water and mixed with rapid stirring, about 18,500 rpm. The interfacial syntheses gave generally better product yields ranging from 20 to 90 % yields while the aqueous solution syntheses gave product yields ranging from 3 to 30%. Yield was studied as a function of stirring rate. For the interfacial systems, percentage yield continued to increase as stirring rate was increased from 9,100 to 21,400 rpm consistent with the increased stirring rate creating fresh surface for polymer to be formed during the reaction window of time. By comparison, for the aqueous solution system, percentage yield remained approximately constant throughout this stirring range consistent with reaction being dependent on only the reactants coming in contact with one another. The activation energy for addition of the Cp_2Ti^{+2} to an amine is about 10 kcal/mol (40 kJ/mol) consistent with this idea. The reaction was also studied as a function of pH. The percentage yield was approximately constant from a pH of 7.4 to 12. This is consistent with the active diamine species being the un-ionized RNH_2.

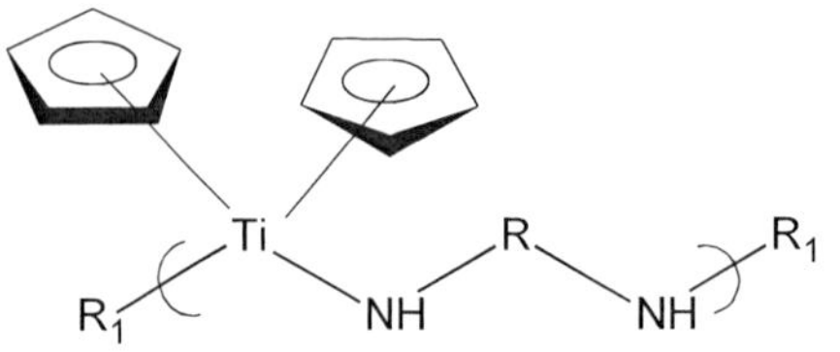

(2)

No product was formed in the absence of added base or buffer consistent with the importance for removal of generated hydrogen chloride and/or importance of the base to assist in the removal of a proton from the amine as it reacts with the metallocene halide. Those derived when organic bases as triethylamine are employed in place of sodium hydroxide show at least some triethylamine end groups, **3**.

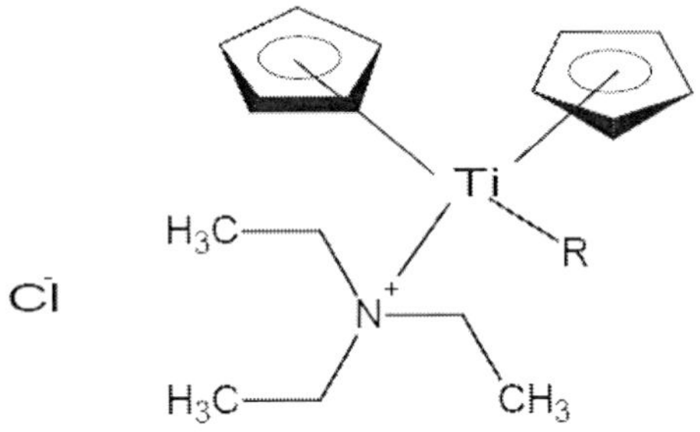

(3)

Both the interfacial and aqueous solution systems showed an increase in yield as the amine to titanocene dichloride ratio increased indicating the importance of having sufficient amine present at the reaction site. Thus, in order to achieve high product yields, relatively high concentrations of diamine should be employed.

Product was also formed using the inverse interfacial system where the diamine is in the organic layer and the metallocene dichloride in the aqueous layer. The metallocene-containing aqueous layers should be made just prior to the reaction to eliminate unwanted reactions such as the formation of titanocene hydroxides and oxides.

Crosslinked products were made using a variety of tetra-amines.[20] Product yields were in the mid-range, about 20 to 40%. The products generally showed better than 60% weight retention to about 900 °C in nitrogen, but in air retention was less with some of the products having mostly titanium dioxide as the final solid. Research was halted on these materials when it was found that some of the tetra-amines were cancer causing agents.

As part of our effort to increase food production in third world countries, Group IVB polymers containing the diamine plant growth hormone kinetin in the backbone were synthesized.[21,22] Of interest is that the zirconocene product was a semiconductor with a (bulk) electrical conductivity of about 600 kilo-ohms, well within the range of being a semiconductor. The product yields were similar for all three Group IVB metallocene dichlorides, all within the range of 55 to 63%. The resulting polymers have a repeat unit as shown in **4** for the zirconocene product.

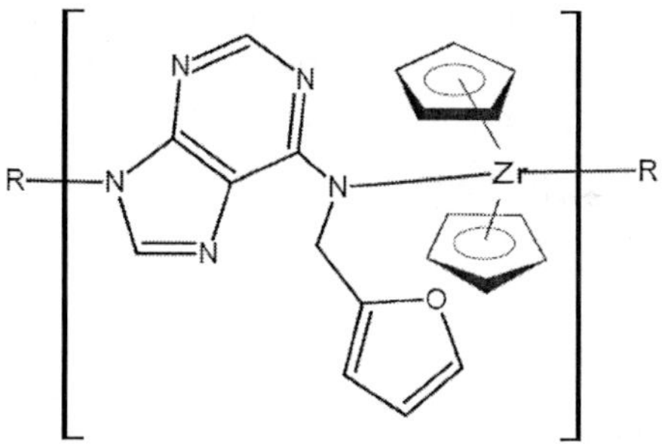

(4)

As part of our effort to assist in the restoration of the Florida Everglades we investigated the ability of talc samples containing ppm of the kinetin-containing polymers to influence sawgrass seed germination. In the wild, sawgrass seed germination is reported to be from zero to 2%. Even in the laboratory the percentage germination of sawgrass seeds was low before our efforts, also near zero. Treatment of randomly selected sawgrass seeds for the control was zero, but for the polymers, in some cases, over 50%. This would allow the Everglades to be re-seeded using a water boat or airplane employing simple casting of treated seeds rather than the present practice of hand planting single seedlings.

We also investigated the ability of these talc-kinetin polymer mixtures to influence the seed germination of damaged and old seeds.[23-25] The polymers generally showed increased germination rates for all seeds tested. Significant increases in germination were found for many of the treatments. The increases were modest for mustard with the largest germination rates found for the higher concentrations (1000 ppm) for the polymers with averages being 32 % (Ti), 34 % (Zr), and 40 % (Hf) in comparison to the control value of 23 %. For the damaged Jagger Wheat, the highest yields were found for kinetin, 10 ppm itself (25 %) and the titanocene-kinetin polymer, 1000 ppm (20 %; compared to a control value of 9 %). These increases represent a doubling in the percentage germination in comparison to the control. The largest differences were found for turnip seeds where all of the treatments gave at least a doubling in the germination rate with several giving about a six to eight fold increase. The highest increase was for the zirconocene-kinetin polymer, 1000 ppm (27 %).

There appeared to be little correlation between the concentration of the treatments and germination rate probably because the precise affect on germination is a complex mixture of factors including polymer degradation rate releasing the kinetin. It is important to note that slight variations in the structure of the cytokinin renders the PGH inactive.[24-29]

Since the polymers containing kinetin exhibit PGH-like activity, the polymers probably release kinetin in an active form as they degrade. While degradation studies have not been performed on the present materials, similar condensation polymers have molecular weight half lives (time required for the average molecular weight to be halved) in HMPA and DMSO solution of days to months but molecular weight studies have not been done on simple exposure of polymers to water. Further, the situation is more complex since microorganism degradation may also occur.

The kinetin-bound polymers generally outperformed kinetin itself. Binding the kinetin with polymers will result in additional cost so that use of kinetin alone might be recommended on this account. Counter, kinetin-containing polymers can be stored in the open under room conditions for over two years without any noticeable change (by IR examination) while the PGHs themselves must be stored under refrigeration.

As noted before, it has been found that any variation in the structure of a plant growth hormone, PGH, will render it inactive.[24-29] Thus, while the structure of the active PGH was not determined it is assumed that the active agent is kinetin itself whenever PGH activity was found. For the polymers this amounts to degradation of the polymer chain releasing kinetin. It is believed that the metallocene moiety degrades to generally non-toxic metal oxide. Thus, the polymers are environmentally “friendly” degrading to environmentally acceptable compounds.

In summary, in almost all cases, polymer-containing PGH treated seed groups gave greater germination percentages in comparison to the control. Thus, such treatment may be advantageous to increasing the germination of older and damaged seeds increasing the number of viable plants from such seed feedstocks. Further, several grams of polymer are sufficient to treat millions of seeds.

In all cases seedling development appeared independent of the particular seed treatment. This is consistent with the favored use of "natural" PGHs incorporated in polymers with respect to their lack of production of negative side effects hopefully avoiding unintended negative consequences.

Recently it was found that a titanocene polyamine was found to increase its conductivity about 10^5 when doped with iodine. [30,31] This is the first example of a traditional condensation polymer whose conductivity was significantly increased through doping.

The polymeric material is derived from the interfacial condensation polymerization between titanocene dichloride and 2-nitro-p-phenylenediamine (Figure 1). This product was initially synthesized by us in 1975 employing both the interfacial and solution polymerization techniques.[19]

The foundation for these efforts was laid by MacDiarmid, Heeger, and Shirakawa (for which they received the Nobel Prize in 2000) and coworkers and much work continues today.[32-39]

We have just begun investigating the results of polymer doping. While there are a number of different doping procedures employed in the production of conducting polymers, the most widely employed is the exposure of the polymeric material, often in a compacted disk, to elemental iodine vapors.[32-39] During our initial efforts we found that one polymer did respond giving relativey large increases in its conductivity when doped with iodine vapors. As noted before, this polymer is obtained from reaction between titanocene dichloride and 2-nitro-p-phenylenediamine (Figure 1).

The titanium atom in titanocene dichloride has a vacant orbital that can accept electrons allowing resonance to occur through it. The polymer can exhibit whole-chain resonance as shown in Figure 2 so its structure is consistent with the general structural criteria for conductivity. The product employed in the current study was synthesized in 45 %. It has a weight average chain length (degree of polymerization) of 310 units.

It was found that simply exposure of a solid pellet of the polymer to iodine vapor allowed the bulk conductivity to increase up to 10^5. It was also found that this process was cyclic. Thus, the pellet was exposed to iodine vapor resulting in an increased electrical conductivity. The pellet was then heated to remove the surface iodine and the conductivity decreased. The same pellet was again exposed to iodine vapor and again the electrical conductivity increased.

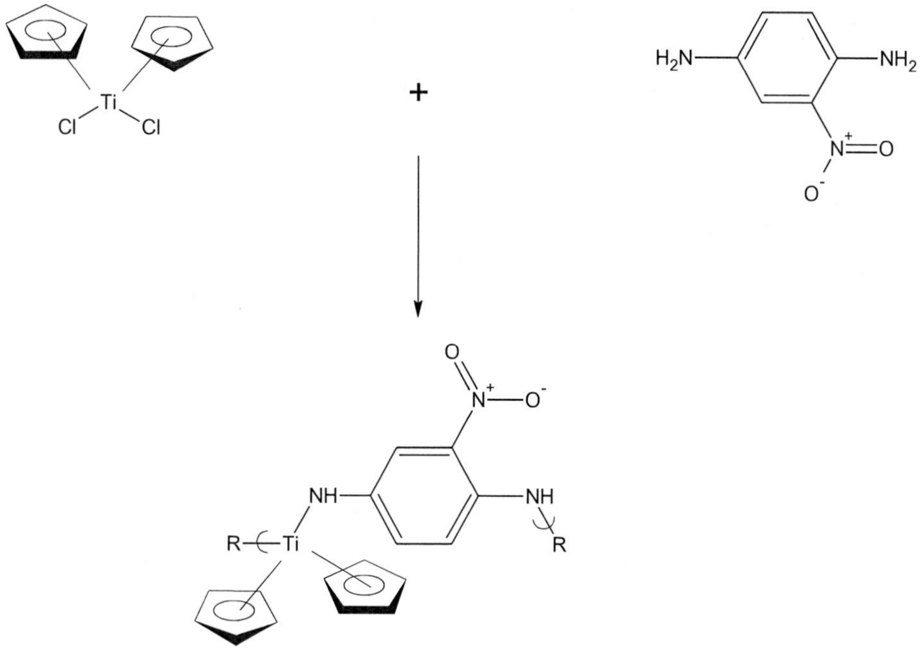

Figure 1. Synthesis of polymer from reaction of titanocene dichloride and 2-nitro-p-phenylenediamine where R is simple chain extension.

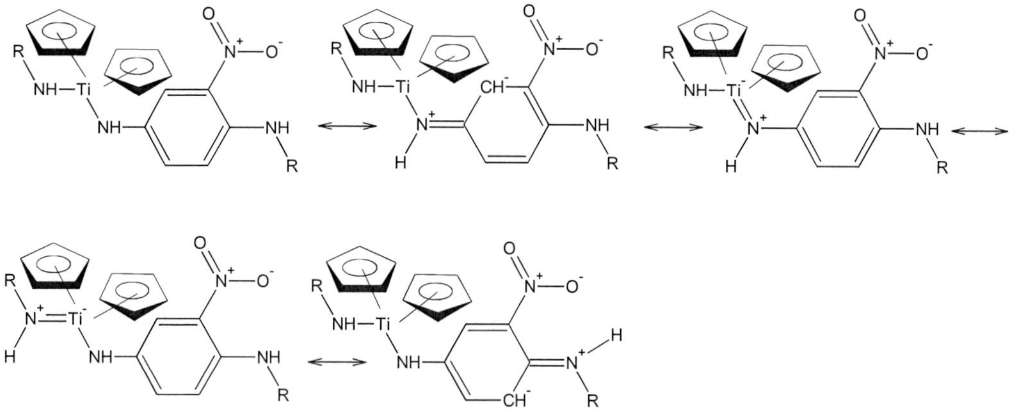

Figure 2. Resonance structures for 2-nitro-p-phenylenediamine unit where R = chain extension.

We also incorporated a number of dyes as part of the backbone of polymers. These polymers are generally referred to as polydyes. Our overall effort was to manage light, to either concentrate light or to disperse it. "Normal" and laser light is of concern to us. One goal is to couple known light harvesting units in a polymer chain with those that disperse light. Group IVB metallocenes are known to "attract" or harvest light, mainly in the visible and UV regions. They are sometimes referred to as UV "sinks" because of their ability to absorb UV radiation without begin degraded. This ability to absorb light and then give it off again has many potential uses including allowing energy absorbed in a coating to be re-emitted at a specific wavelength that would allow the coating to cure at greater depths and more rapidly. The absorption of light energy by a coating containing the polydye may also be part of an overall effort to harvest radiation to contribute to the heating and supplying of energy (including electrical energy) to buildings and dwellings.

Another attraction for including dyes into polymers is the (potential) permanent, non-leaching nature of such polymer-containing dyes allowing a material that contains this polydye to remain colored for longer times. Even within solids, such as plastics, migration and loss of colorants occurs, especially over a long period of time. The ability to retain color or other important additive property is especially important in applications such as in marine coatings where leaching and subsequent loss is aided by the ever present aqueous environment.

A number of dyes were examined including those derived from diamines such as nigrosine, **5**.[40-49]

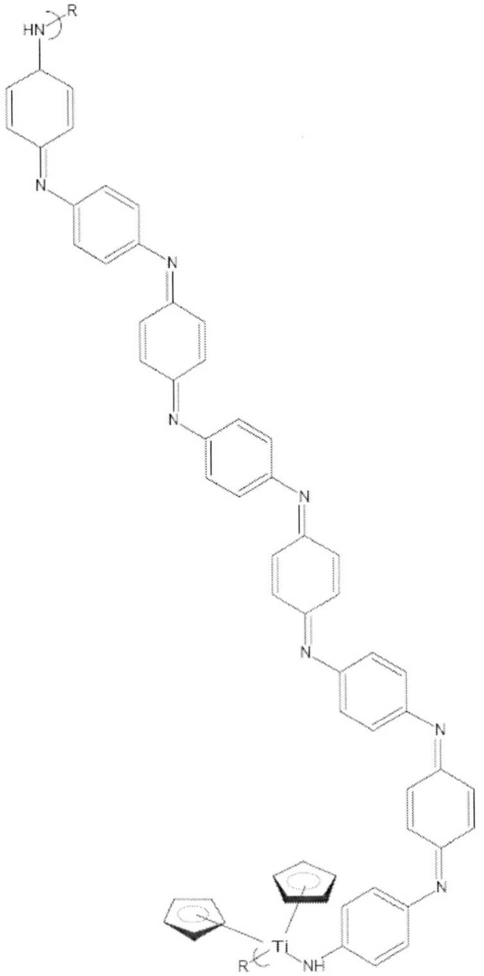

HN R N N N N N N N N Ti R NH

(5)

The polydyes are generally soluble in dipolar aprotic liquids such as hexamethylphosphorus triamide (HMPA), N,N'-dimethylformamide (DMF), dimethyl sulfoxide (DMSO), and triethylphosphate (TEP). Polydye solutions were used to impregnate a number of items including plastics (polyethylene, polycarbonate, polypropylene, poly(vinyl chloride), polystyrene, nylon 6,6, SAN-styrene copolymer) , cloth (cotton, denim, and 50-50% polyester fabric), paper, films, and coatings.

In one set of experiments, comparison was made with simple solutions containing the dyes themselves. After several days to months, the color of the dye solutions containing only the dyes themselves were greatly decreased. By comparison, the polydyes retained decent color for a year under fluorescent lighting. Exposure to outdoor light hastened deterioration of the color of both the dye and polydye solutions. Thus, incorporation into a polymer appears to lock in the dye structure and gives it some added stability to ordinary in-door light. In the dark polydye, solutions have remained fluorescent and colored now for about 25 years.[49]

Studies were undertaken using an Argon Ion laser operated at 514.5 nm (visible-green) as the high energy source. Optically transparent films were cast. The presence of polydye in the samples caused significantly less (almost ten-times shorter) "burn-through" times compared with those containing an equal amount of the dye itself.[48] In this study, the polydyes acted to concentrate the radiation dramatically decreasing the "burn-though" times. By comparison, similar samples except exposed to energy from a carbon dioxide laser operating in the infrared region (1.06 microns; 1060 nm).[40] Here burn-through times were significantly greater. Under these experimental conditions, the polydyes acted as dispersing agents. The films contained from 10 to 100 ppm polydye.

In another set of experiments using the carbon dioxide lazer, wood was coated with latex paint containing about 200 ppm polydye. The wood was exposed to laser light in the IR region. In comparison to non-impregnated latex coatings, the polydye-containing samples exhibited equal and greater burn-through times. Similar results were found for plastics doped with polydyes.[40,49]

Behavior was dramatically different depending on the range of employed radiation. Thus, in the IR region the polydyes act as dispersing agents protecting the material that is coated with it or that contains it. In the visible-UV region the polydye acts as a concentrating agent, concentrating the effect of the radiation.

Each tendency offers industrially important behavior. For instance, some chip fabrication and construction might be enhanced through addition of some polydye material that allows fabrication through use of lasers employing lower energies. The use of lower energies would result in lower temperatures that, in turn, would protect the chip from thermal induced deterioration. A similar advantage is achieved when cutting or etching occurs. Counter, in some applications greater stability is needed to resist potentially damaging laser radiation.

Ruthenium Containing Polyamines

We have synthesized a number of ruthenium-containing polymers, mostly in a search for converting solar energy or other light energy into usable energy under the heading of solar energy conversion. It is of interest to note that our work in this area involves two different types of reactions, the coordination reactions involving reactions with diamines and condensation reactions involving thiols. While the present paper focuses on condensation polymers, we will briefly describe efforts involving the synthesis of ruthenium-containing polyamines employing the coordination process.

Keen, Salmon, and Meyer [50] reported the synthesis of monomeric amines from reaction of ruthenium(II) bis-2,2'-bipyridine dichloride, DBR, with diamines where the reaction is a coordination reaction rather than the condensation reaction that occurres with the thiols.

$$RNH_2 + Ru(bpy)_2Cl_2 \longrightarrow (RNH_2)_2Ru(bpy)_2Cl_2 \qquad \text{Equation 3.}$$

Reaction with diamines gave the analogous polymeric products. [51-55]

$$H_2N\text{-}R\text{-}NH_2 + Ru(bpy)_2Cl_2 \longrightarrow \text{-}(\text{-}H_2N\text{-}R\text{-}NH_2\text{-}Ru(bpy)_2\text{-})_n\text{-}\ ,\ 2Cl \qquad \text{Equation 4.}$$

These efforts have two major aims; first, modification of the accepting and emitting energies of the ruthenium center, and second, to assist in attracting light that can be subsequently harvested by the ruthenium center, thus increasing the converted quantum yield of harvest light. The current effort is aimed at incorporating dyes into the ruthenium-containing polymers.[51,52]

In one series, products were formed from reaction of cis-dichloro-bis(2'2-bipyridine)ruthenium II, $Ru(bipy)_2Cl_2$, with the dye suspended in an 80 % by volume methanol aqueous solution. This mixture was refluxed on a steam bath for about six hours, after which the methanol was allowed to evaporate. The reaction mixture was cooled over night. The solution was washed with benzene removing unreacted diamine and washed with chloroform removing unreacted $Ru(bipy)_2Cl_2$. The remaining liquid was removed under vacuum giving the product as a solid, generally in good yield.

The product from $Ru(bipy)_2Cl_2$ and 6-butoxy-2,6-diamino-3,3'-azodipyridine, diazopy, **6**, is produced in good yield, 89%, and is soluble in DMSO, methanol, ethanol, and slightly soluble in chloroform. It is insoluble in benzene, carbon tetrachloride, acetone, and hexane. It has a molecular weight of 1.4×10^5 Daltons which corresponds to a degree of polymerization of about 180.

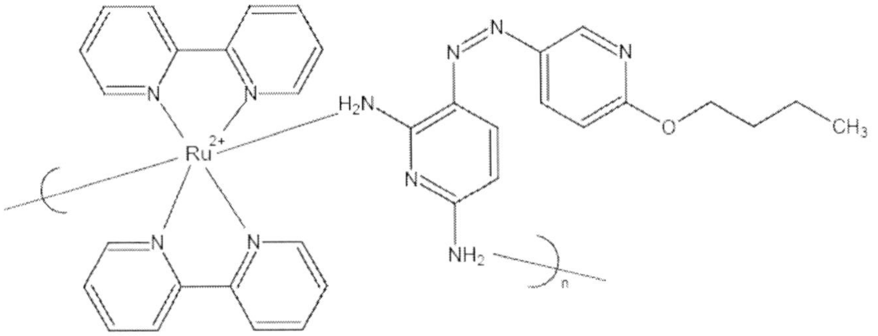

(6)

Here we describe only the UV-VIS results since these most closely align with the use of such materials in light harvesting. Spectra were obtained in DMSO.

The ruthenium starting material has an octahedral geometry. The ruthenium monomer exhibits bands at 557, 379, and 300 nm with the 300 nm band being the lambda maximum band, which is in the middle of the ultraviolet region. These bands are associated with eg to t_2g d-electron transition. Substitution of the chloro ligands by amines should cause a change in the energy for d-electrons to undergo an eg to t_2g transition.

The diazopy dye itself has two bands at 291 nm and the lambda maximum band is at 454 nm. The polymer product shows a number bands with the bands appearing at 495, 437, 355, 328, 295, 266, and 252 nm. There is a broad band on the higher side of 500 nm that probably corresponds to the 557 nm band found in the ruthenium monomer. The lambda maximum band for the dye that appeared at 454 nm is now at 437 nm, still in the blue region of visible light. The lambda maximum band for the ruthenium monomer that appeared at 300 nm is now at 295 nm for the polymer.

For the product of $Ru(bipy)_2Cl_2$ and diazopy all the bands from the two reactants are essentially the same except for the lambda maximum band for the dye at 454 nm which is missing. However, the bands are broader for the polymer and the 454 nm band is now probably lost within this broadening caused by the $Ru(bipy)_2$ moiety. The product also contains a large peak at 295 nm. The 295 nm band is the lambda maximum for the product as is the 300 nm for the starting ruthenium material. There are also broad peaks at 355 nm and 495 nm with a less intense peak at 252 nm. Interestingly, the lambda maximum for the dye itself at 454 nm now corresponds to a much smaller peak at 437 nm.

A similar product was formed from the reaction of $Ru(bipy)_2Cl_2$ and N,N'-bis(3-aminophenyl)-3,4,9,10-perylenetetracarboxylic diimide, **7**. This product was formed in 85 % yield and had a DP of 33,000 by light scattering photometry.

The broadness of the spectral bands may be viewed as positive with respect to the polymer accepting a wide range of wavelengths of light. Emission spectra were not taken of the materials.

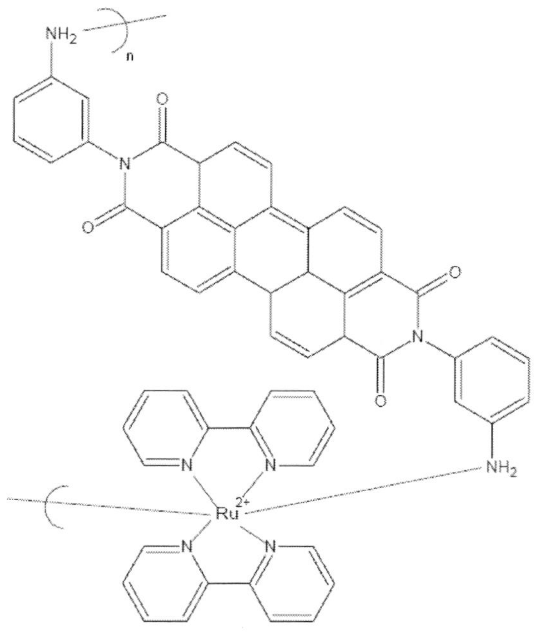

(7)

Similar products will continue to be studied as the search for inexpensive energy continues.

Main Group Polyamines

Organosilicone Polyamines

Organosilicone polyamines were initially synthesized by us in 1969 (**8**). These materials were generally of low molecular weight and often were tar-like rather than solid. This is probably due to the high rate of hydrolysis of organosilane halides in comparison to organotin halides with hydrolysis effectively terminating chain growth.

$$Cl-SiR_2-Cl \;+\; H_2N-R_2-NH_2 \longrightarrow \sim R_1-SiR_2-NH-R_2-NH-R_1\sim \qquad (8)$$

Along with the traditional diamines, low molecular weight products were formed from reaction with hydrazine (**9**). [56]

$$Cl-SiR_2-Cl \;+\; H_2N-NH_2 \longrightarrow \sim R_1-SiR_2-NH-NH-R_1\sim \qquad (9)$$

A modified interfacial system was developed employing 2,5-hexanedione or acetonitrile in place of the usual aqueous phase in an attempt to reduce hydrolysis. Also, various bases were employed along with sodium hydroxide. The best results were obtained when employing triethylamine as the added base. [57]

Other hydrazines were also employed employing the non-aqueous interfacial system. [58] Yields to about 100% were obtained but similar to other organosilicone polyamines, yield generally decreased as the stirring time increases. This is consistent with a lack of good aqueous stability for the Si-N bond. Chain length generally increased with stirring time probably indicating that the first polymer chains to be degraded were the lower mass chains leaving the higher mass chains resulting in an apparent increase in chain length.

Additional studies were undertaken including developing other polymerization systems and investigating reaction variables. [59,60]

Organotin-Containing Polyamines

The first organotin compound was prepared by Sir Edward Frankland in 1849.[61] The topic of organotin polymers has been reviewed elsewhere.[62,63]

There are a larger number of organotin compounds in commercial use than for any other metal. [64] This has resulted in an increase in worldwide production of organotin compounds over the last few decades. This production exceeded fifty thousand tons in 1992 and accounts for about 7% of the tin metal used. [64]

Organotin polymers are important industrially serving as poly(vinyl chloride), PVC, heat stabilizers, in film for food packaging, and in PVC articles.[62,63] They are also used as antiseptic, antifouling, and anti-mildew agents in industry and agriculture and as additives in paint formulations. Films have excellent transparency.

PVC is unstable under exposure to light and heat resulting in discoloration and embrittlement. In the early 1940s it was found that this degradation could be prevented by addition of certain organotin derivatives. Presently, about 70% of the commercial organotin compounds are employed as PVC stabilizers in the form of mono and dialkyltin derivatives. Today, many of these stabilizers are based on organotin polymers made by Carraher and co-workers. [63]

Much of the more recent drive towards inclusion of organotin into polymers is the result of federal legislation that prohibits use of so-called unbound or monomeric organotin compounds in paints and coatings. Organotin monomeric compounds were widely used as antifouling and antimold agents but through migration they inhibited and killed nearby plant and animal life. (Chemically) bound organotin was permitted in the legislation. Thus, there is activity to create non-migrating chemically bound organotin materials, namely polymers that contain organotin moieties.

While much of our effort involves the synthesis of organotin polyamines from reaction with diamines, organotin polymers have been synthesized from reaction with preformed amine-containing polymers. Using polyethyleneimine, the products will be linear as the monohalo organotin is reacted with the polyethyleneimine, **10**, but they will be cross-linked when dihaloorganotin reactants with two functional groups are employed, **11**.

$$\begin{array}{c} SnR_3 \\ | \\ \text{-(-}CH_2\text{-}CH_2\text{-N-)}_n\text{-} \end{array} \qquad (10)$$

and

$$\begin{array}{c} | \\ SnR_2 \\ | \\ \text{-(}CH_2\text{-}CH_2\text{-N-)}_n \end{array} \qquad (11)$$

As in the case with the Group IVB metallocenes, we also synthesized a number of plant growth-containing polymers including polyamines derived from kinetin. The repeat unit for the product derived from dimethyltin dichloride is given below, **12**. The product is then an organotin polyamine. The use of these materials allowed the sawgrass seed germination rate to increase from the 0-2% range to over 20%. We have now systems that will give sawgrass germination rates of over 60% that makes wholesale seed distribution by air boat or airplane a practical way of seeding large areas in need of sawgrass restoration.[65-68] The organotin-containing products offer an advantage over the metallocene product in that they are resistant to degradation of many molds, yeasts, and bacteria because of the presence of the organotin moiety.

These products also exhibit an ability to increase the germination of damaged and old food crop seeds.

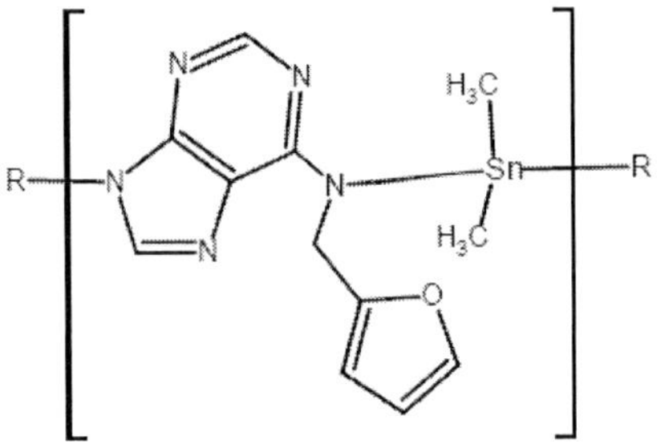

(12)

Because of the increased worldwide production or organotin compounds for commercial use considerable amounts of organotins have entered various ecosystems. Inorganic tin compounds are often said to be non-toxic, but in fact their toxicity ranges from being non-toxic to moderately toxic. Organotin compounds offer varied toxicities from being mildly toxic to highly toxic with most of the compounds falling in the mildly to moderately toxic range. The toxicity and ecosystem relationships have been recently reviewed. [64]

While organotin-containing polymers offer a wide variety of biological activities we will focus on their use in the inhibition of bacterial and yeast at this juncture of the paper.

The general microorganism biological activities of organotin compounds has been studied since the 1950s. As noted above, generally inorganic tin compounds are non-toxic or only slightly toxic towards mammals, insects, bacteria, and fungi whereas organotin compounds show varying biological activities. For alkylorganostannanes, toxicity varies depending on the alkyl group. In general an increase in the alkyl chain length gives a decrease in toxicity.[69] The particular pattern for methyl, ethyl, propyl, and butyl varies with test organism. For tri-n-alkyltin acetates, the methyl organotin compounds are the most active for insects and mammals; for fungi and bacteria the propyl and butyl compounds are the most active. Further, activity decreases as the number of alkyl groups decreases as follows $R_4Sn > R_3Sn > R_2Sn > RSn > Sn$ all for tin IV compounds. For most cases our findings are consistent with these general trends and long alkyl chains such as n-octyl are generally biologically inactive with respect to bacterial inhibition.[70-94]

The organotin materials are suitable for treatment of infections (here mainly topical), contaminated sites, use as a preventative agent, and in the treatment of water sources. The organotin-containing drugs are rapidly (generally within 30 seconds) synthesized in good yield employing readily available reactants. Thus, ready availability on a gram to tons scale of target drugs is achievable. They can be used internally or topically as additives to creams, cleaning detergents and soaps, coatings (paints), plastics, paper, etc. [89] They can be generally be handled without need for gloves or other protective ware but it is always wise to be cautions because the individual toxicity of individual products varies with the person and product. Further, if the reactants have not been adequately removed, they can present adverse effects. These polymers typically have shelf-lives in excess of several years.

Following are microorganisms that have been successfully inhibited by condensation organotin-containing polymers including organotin polyamines.

E. coli,
B. catarrhalis,
E. aerogenes,
B. subtilis,
S. epidermidis,
N. mucosa,

K. pneumoniae,
A. flavus,
A. fumagatus,
Trichoderma reesei,
P. aeruginosa,
C. albicans,
Staph MRSA,
B. subtilis,
A. calcoacetius,
A. niger,
Penicillin sp.,
Chaetomium globosum,
S. aireis,
T. mentagrophytes,
S. cerevisiae,
Al. faecalis

Most of these products have been incorporated into paper, plastics, textiles, and the like with only some loss in biological activity.[89] We will now move to the use of organotin polyamines as potent anticancer drugs. [74,74,96-100] Recently we synthesized a wide variety of organotin polymers derived from 4,6-diaminopyrimidines (Figure 3).[101]

The molecular weights ranged from 3.5×10^4 to 3.7×10^6 and product yields from 47 to 88 %. The polyamines were synthesized for two main reasons. The first reason was to determine the biological activities of the products. The second was to determine the electrical properties of the polyamines. Because of the variety of products formed it should be possible to relate such factors as electronic nature of the pyrimidine, chain length and steric nature of the polymers to their electronic and biological activity. In general, the products exhibit good inhibition of cancer cell lines including those derived from human lung, bone, breast, prostrate, and colon cancers. Following is a brief discussion of these results.

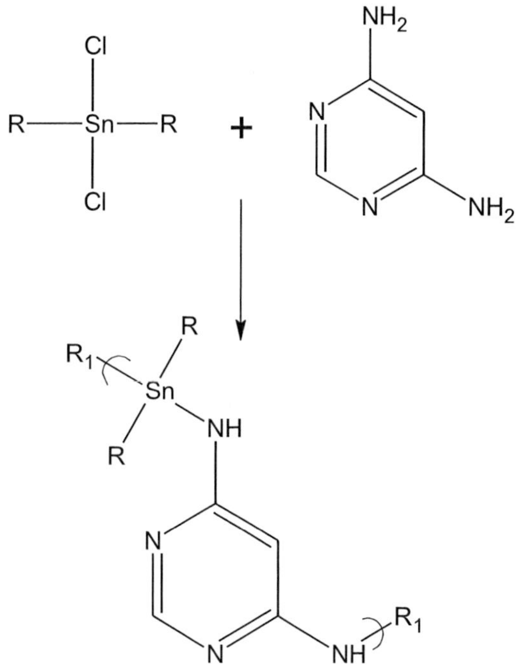

Figure 3. Reaction of organotin dichlorides with 4,6-diaminopyrimidine.

Different measures are employed in the evaluation of cell line results. Here we use the two most widely employed- growth inhibition 50%, GI_{50}, values which is the lowest concentration where growth is inhibited by 50% and the Chemotherapeutic Index, CI_{50} which is a measure of the amount needed to inhibit 50% cell growth, GI_{50}, for the normal cell line, here WI-38 cell line, divided by the amount needed to inhibit 50% cell growth for one of the cancer cell lines. It is to be noted that different researchers generally emphasize one of these measures over the other with neither measure universally accepted. Thus, results from both of these measures are presented.

One reason for synthesizing this series of organotin polymers is to investigate if a clear trend can be established with respect to the electronic, or some other, nature of the pyrimidine. Two thiol-containing pyrimidines were employed because of the need for sulfur in the body and the general lack of sulfur-containing sources. This may encourage inclusion of the thiol pyrimidine units at important junctures in the cell growth cycle. Substitution on aromatic rings, in general, is consistent with groups such as hydroxyl, aliphatic, aromatic, and thiol releasing electrons to the ring resulting in a site that is electropositive. Conversely, the presence of groups such as nitro, nitroso, and halides withdraw electrons from the ring resulting in an electronegative environment for the pyrimidine ring. Many of the pyrimidines used in this study contain both electron withdrawing and electron donating substituents. The GI_{50} concentration values for the polymers are in the same general range of 2 to 0.04 micrograms/mL regardless of the identity of the pyrimidine. All of the polyamines offer reasonable inhibition of all of the cancer test cell lines and are in the same general range as cisplatin, the most widely employed chemo drug.

The second measure is the 50% chemotherapeutic index, CI_{50}. As noted before, the chemotherapeutic index is the concentration of the compound that inhibits the growth of the healthy cell by 50% divided by the concentration of the compound that inhibits the growth of the cancer cell by 50%. Larger values are desired since they indicate that a larger concentration is required to inhibit the healthy cells in comparison to the cancer cells or stated in another way, larger values indicate some preference for inhibiting the cancer cells in preference to the normal cells. In general, CI_{50} values larger than 2 are considered significant.

Both, the standard, dibutyltin dichloride, and all but two of the polymers, show at least one instance of a CI_{50} value of 2 and greater. Some show large, about 50% of the compounds, values much greater than 2. Also of interest is the behavior towards 2RA. 2RA is a fully transformed WI-38 and this couple is often taken as an indicator of the relative toxicities of drugs to a healthy cell compared with a cancer cell. Seven of the nine polymers exhibited CI_{50} values greater than one for the ratio of WI-38/2RA, and four of the nine polymers displayed CI_{50} greater than 2 which is consistent with the polymers being more toxic to the cancer cell line derived from the healthy cell line than the healthy cell line itself. Of the three polymers that generally exhibited the lowest GI_{50} values only one showed more than one instance where the CI_{50} values were equal to and greater than 2. Thus, there is little correlation between the relative toxicities measured by these two common measures. Further, there appears to be little correlation between the electronic nature of the pyrimidine and either GI or CI values. It is of interest that both of the pyrimidines that contain thiol groups exhibit low GI and high CI values.

The results also do not show a relationship between chain length or steric nature and ability to inhibit cancer cell growth.

Pancreatic cancer afflicts close to 32,000 individuals each year in the United States and 168,000 worldwide, and nearly all patients die from the ravages of their disease within 3 to 6 months after detection. It is the fourth leading cause of cancer death. Treatment of pancreatic cancer is rarely successful as this disease typically metastasizes prior to detection. Current therapies consist of surgery and, possibly, radiation and chemotherapy. Standard chemotherapy for patients with locally contained cancer includes gemcitabine. Gemcitabine has been demonstrated to improve the quality of life through better pain control, adequate performance status, decreased analgesic consumption, shrinkage of tumor, and prolonged survival. Radiation therapy is usually ineffective except as an adjunct to chemotherapy or as a palliative measure. There is no effective chemotherapy for metastasized pancreatic cancer. 5-Fluorouracil (5-FU) is the most widely used agent for single drug therapy in the treatment of pancreatic cancer. 5-FU is an antimetabolite that interferes with cellular processes essential for cell division and cell growth. As a result, 5-FU has the greatest effect on rapidly growing cells.[102-104]

Considering the lack of substantial impact on survival of any chemotherapy regimens, all patients with pancreatic cancer should be considered for enrollment in clinical trials. If this is not possible, gemcitabine appears to be emerging as the standard treatment due to its apparent favorable impact on disease-related symptoms, such as pain control and performance status.[105-108]

Three of the 4-6-diamiopyrimidine polymers exhibited CI_{50} values of 20 and greater towards two strains of pancreatic cancer, APSC-1 and PANC-1. [109] There polymers are derived from reaction of dibutyltin dichloride and 4,6-diamino-2-mercaptopyrimidine, 2-chloro-4,6-diaminopyrimidine, and 4,6-diamino-2-hydroxypyrimidine.

Some of the organotin polypyrimidines also exhibit excellent inhibition of both the HSV-1 virus and the Vaccinia virus. [110] The HSV-1 or herpes simplex virus is responsible for at least 45 million infections in the US or one of every five adolescents and adults. The Vaccinia virus is the viral strain responsible for small pox and is considered one of the viruses that might be employed in a viral terror attack. It was also found that these pyrimidine polyamines exhibit, as solids, decent inhibition of a variety of bacteria and yeasts. [111]

The 4-6-diaminopyrimidine polymers have structures similar to polypyrrole and exhibit whole or entire chain resonance. Even so, their inherent electrical conductivities were all within the non-conductor to semi-conductor range with no relationship between conductivity and electrical nature of the pyrimidine, chain length, or steric hindrance.[112]

GROUP V POLYAMINES

Doak and Freedman [113] synthesized the monomeric organoantimony amines, **13**, from reaction of organoantimoney dihalides with monoamines.

$$R_3SbCl_2 + R_1\text{—}NH_2 \longrightarrow R_1NH\text{—}SbR_3\text{—}NHR_1 \qquad (13)$$

The analogous polyamines,**14**, were synthesized employing the diamine.[114-117]

$$R_2SbCl_2 \text{(R)} + H_2N{-}R_1{-}NH_2 \longrightarrow \text{-}[Sb(R)_3{-}NH{-}R_1{-}NH]\text{-} \quad (14)$$

Yields ranged from about 10 to over 90%. A wide range of diamines were employed including simple aliphatic diamines such as 1,6-diaminohexane, simple aromatic diamines such as p-phenylenediamine, to more complex diamines such as adenine, 2,6-diamino-8-purinol, 4,4'-diaminodiphenylsulfon, Zineb, and 2,4-diamino-5(3,4-dimethoxybenzil) pyrimidine.

The molecular weights of the products were within the range of 10^3 to 10^6 Da. The product of triphenylantimony dichloride and 4,6-diamino-2-mercapto pyrimidine has a molecular weight of 5.7 x 10^5 corresponding to a DP of about 1,200. The product from triphenylantimony dichloride and 4,4-diaminodiphenylsulfon is only oligomeric with a molecular weight of 3 x 10^3 Da corresponding to a DP of 10.

The products were tested for their ability to inhibit a wide range of bacteria, fungi, and yeast. [114] The ability to inhibit ranged from essentially no inhibition against any test organism for the product from adenine and triphenylantimony dichloride to inhibition of all ten test organisms for the 4,6-diamino-2-methyl-5-nitrosopyrimidine and triphenylantimony dichloride product.

The products were also tested for their ability to inhibit a number of cancer cell lines including BHK-21, L929, and HeLa cells.[114] The products generally exhibited decent cell inhibition with GI_{50} values in the range of 5 micrograms/mL for the HeLa cells.

A number of other triphenylantimony dichloride-derived products were recently tested for their ability to inhibit Balb 3T3 cells as a measure of their ability as potential anticancer drugs. [116,117] This included those derived from cobalticinium-1,1'-dicarboxylic acid nitrate and thiopyrimidine. These materials showed GI_{50} values of about 10 micrograms/mL. By comparison, the most widely used anticancer drug, cisplatin, has a GI_{50} against the same strain of 3T3 cells of about 4 micrograms/mL. Thus, the antimony-containing polymers exhibit an ability similar to that of cisplatin against 3T3 cells.

The effect of use of phase transfer agents and crown ethers was studied for the synthesis of a number of antimony polyamines. [116] Comparison was made with the use of sodium hydroxide alone as the added base.

Results were varied with some added catalysts resulting in an increase in product yield and chain length, others giving a lower yield and chain length, and still others showing no change in product yield and chain length. For instance, the use of tetraphenyl phosphonium iodide gave an increase from 35 to 45% in product yield and a threefold increase in molecular weight for the reaction of 2,6-diamino-8-purinol and triphenylantimony dichride. By comparison, the same phase transfer catalyst showed a decrease in yield from 38 to 28% and a four fold decrease in chain length for the reaction of 2,5-dichloro-p-phenylenediamine and triphenylantimony dichloride in comparison to the simple sodium hydroxide system.

A number of arsenic polypyrimidines have been synthesized (**15**). These polymers are good inhibitors of a number of bacteria.[118, 119]

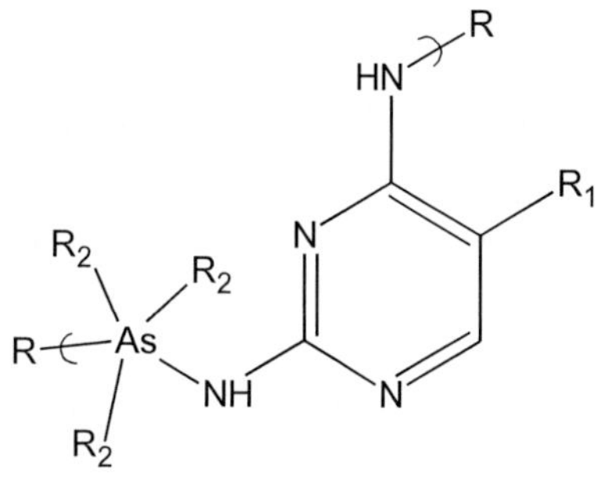

(15)

FUTURE

While many metal-containing polyamines have been synthesized, many more are easily synthesized employing techniques already developed. It is important to remember that the polymers described in this report are easily and readily synthesized from commercially available reactants employing industrially utilized sequences. Further, synthesizes are generally rapid occurring in less than one minute. Thus, commercialization of any particular material is straightforward.

The concept of incorporating ruthenium-containing units within polymers for attracting solar energy is reasonable and additional work is merited towards this end. Unlike many competing efforts, the synthesis of these polymers is straight forward employing commercially available reactants. Further, the products offer reasonable solubility in a variety of solvents. Water solubility should be possible though the use of poly(ethylene glycol) solubilizing arms such as those employed to cause a number of otherwise insoluble conducting polymers to become water soluble. This approach is true for all of the products described in this paper.

As we continue to move towards nano parts, industrial use of polydyes to control ordinary and laser light also deserves further consideration since it requires the use of only minute, ppm, amounts of the polydye to bring about profound changes in the stability of the polydye-containing material. This is particularly true for the electronic industry with the incorporation or coating of parts can allow more precise fabrication of parts.

While much of the work dealing with cancer has focused on the use of organotin-containing polymers, the metallocene-containing polymers have also been found to offer decent inhibition of a variety of cancer cell lines. Both the organotin and metallocene-containing polymers generally have low toxicities in comparison to the monomeric drugs currently employed in chemotherapy today. Further, the biological testing results are consistent with the metallocene and organotin monomeric products acting at different junctures to inhibit cancer cell growth. In turn, the metallocene and organotin products appear to act at different junctures in comparison to other currently employed chemo drugs. Thus, mixtures containing either or both the metallocene and organotin drugs and currently employed drugs should have a positive effect through intersecting cancer cell growth at different junctures. The advantageous of employing polymeric anticancer drugs is well known.[for instance 120]

REFERENCES

[1] P. W. Morgan, Condensation Polymers by Interfacial and Solution Methods, *Interscience*, NY, 1965.
[2] F. Millich, C. Carraher, Interfacial Synthesis. Vol I. *Fundamentals*, Dekker, NY., 1977.
[3] F. Millich, C.Carraher, Interfacial Synthesis. Vol. II. *Polymer Applications and Technology*. Dekker, NY. 1977.
[4] C. Carraher, J. Preston, Interfacial Synthesis Vol. III, *Recent Advances*, Dekker, NY, 1982.
[5] T. Kealy, P. Pauson, *Nature,* 168, 1039, 1951).
[6] S. A. Miller, J. Tebboth, J. Tremaine, *J. Chem. Soc.*, 632 (1952).
[7] F. A. Cotton, G. Wilkinson, *Advanced Inorganic Chemistry*, Wiley, NY, 1988.
[8] E. Neuse, H. Rosenberg, *Metallocene Polymers*, Dekker, NY, 1979.
[9] P. Wailes, R. Coutts, H. Weigold, *Organometallic Chemistry of Titanium, Zirconium, and Hafnium,* Academic Press, NY, 1974.
[10] G. Wilkinson, F. Stone, E. Abel, *Comprehensive Organometallic Chemistry*, Pergamon, Oxford, UK, 1982.
[11] E. W. Neuse, "Metallocene Polymers", in *Encyclopedia of Polymer Science and Technology,* Vol 8 Wiley-Interscience, NY, 1968.
[12] A. Alaa Abd-El-Aziz, C. Carraher, C. Pittman, J. Sheats, M. Zeldin, Macromolecules Containing Metal and Metal-Like Elements, Vol. 2. *Organoiron Polymers*, Wiley, Hoboken, 2004.
[13] C. Carraher, unpublished.
[14] C. Carraher, *Polymer Chemistry*, 6th Edition, Dekker, NY, 2003.
[15] R. H. Grubbs, C. Lau, R. Curkier, C. Brubaker, *J. Amer. Chem. Soc.*, 99 (1977).
[16] C. Pittman, C. Carraher, J. Reynolds, *Encyclopedia of Polymer Science and Engineering,* 2nd Ed., Wiley, 1987, Vol.10.
[17] C. Carraher, S. Jorgensen, *Polym. Prepr.*, 16(1), 671 (1975).
[18] C. Carraher, P. Lessek, *Europ. Polym. J.*, 8, 1339 (1972).
[19] C. Carraher, S. Jorgensen, *J. Polym. Sci., Polym. Chem. Ed.*, 16, 2965 (1978).
[20] C. Carraher, R. Feiffer, P. Fullenkamp, *J. Macromol. Sci.-Chem.*, A10, 1221 (1976).
[21] C. Carraher, D. Chamely, *Polym. Mater. Sci. Eng.*, 85, 358 (2001).
[22] C. Carraher, D. Chamely, S. Carraher, H. Stewart, *Polym. Mater. Sci. Eng.*, 85, 381 (2001).
[23] C. Carraher, D. Chamely, S. Carraher, H. Stewart, W. Learned, J. Helmy, K. Abby, *Polym. Mater. Sci. Eng.*, 85, 375 (2001).
[24] J. Peterson, C. Carraher, A. Salamone, A. M. Francis, *Polym. Mater. Sci. Eng.*, 81, 149 (1999).
[25] C. Carraher, D. Chamely-Wilk, S. Carraher, G. Barot, H. Stewart, W. Learned, *J. Polym. Mater.*, 24, 149 (2007).
[26] J. P. Metraus, *Plant Hormones and Their Role in Plant Growth and Development*, P. Davies, Ed., pp 296-317, Kluwer, Boston, 1987.
[27] M. Guerinot and D. Salt, *Plant Physiology*, 125, 164 (2001).
[28] C. Somerville and D. Bonetta, *Plant Physiology*, 125, 168 (2001).
[29] A. Trewavas, *Plant Psychology*, 125, 174 (2001).
[30] A. Battin, C. Carraher, *Polym. Mater Sci. Eng.*, 99, 368 and 371 (2008).

[31] A. Battin, C. Carraher, *J. Polym. Mater.*, 25, 23 (2008).
[32] C. K. Chiang, Y. W. Park, A. J. Heeger, A. J. Shirakawa, E. J. Louis, A. G. MacDiarmid, *J. Chem. Phys.*, 69, 5098 (1978).
[33] H. Shirakawa, L. J. Edwin, A. G. MacDiarmid, C. K. Chiang, A. Heeger, *J. Chem. Soc. Chem. Comm.*, 16, 578 (1977).
[34] C. K. Chiang, M. A. Druy, S. C. Gau, A. Heeger, E. J. Louis, A. G. MacDiarmid, Y. W. Park, H. Shirakawa, *J. Am. Chem. Soc.*, 100, 1013 (1978).
[35] C. K. Chiang, C. R. Fincher, Y. W. Park, A. J. Heeger, H. Shirakawa, E. J. Louis, S. Gau, A. G. MacDiarmid, G. Alan, *Polym Comm.*, 31, 1098 (1977).
[36] S. Pekker, A. Morin, F. Beniere, *Polym Comm.*, 31, 75 (1990).
[37] A. J. Epstein, H. Rommelmann, M. Abkowitz, H. W. Gibson, *Mol. Crystals Liquid Crystals,* 77, 81 (1981).
[38] C. Mathai, S. Saravana, M. Anantharaman, S. Venkitachalam, S. Jayalekshmi, *J. Physics D: Appl. Physics*, 35, 2206 (2002).
[39] U. Sajeev, C. Mathai, S. Saravana, R. Ashokan, S. Venkatachalam, M. Anantharaman, *Bull. Mater. Sci.*, 29, 159 (2006).
[40] C. Carraher, V. Foster, R. Linville, D. Stevison, R. Venkatachalam, *Adhesives, Sealants, and Coatings for Space and Harsh Environments*, (L-H. Lee, Ed.) Plenum, NY, 1988.
[41] C. Carraher, R. Schwarz, M. Schwarz, J. Schroeder, *Org. Coat. Plast. Chem.*, 42, 23, (1980).
[42] C. Carraher, R. Schwarz, M. Schwarz, J. Schroeder, *Org. Coat. Plast. Chem.*, 43, 798 (1981).
[43] C. Carraher, J. Kloss, *Polym. Mater. Sci. Eng.*, 64, 229 (1991).
[44] C. Carraher, R. Schwarz, J. Schroeder, M. Schwarz, *J. Macromol. Sci.-Chem.*, A15(5), 773 (1981).
[45] C. Carraher, R. Schwarz, J. Schroeder, M. Schwarz, Interfacial Synthesis, Vol. III, *Recent Advances* (C. Carraher, J. Preston, Eds.,), Dekker, NY, 1982.
[46] C. Carraher, V. Foster, R. Linville, D. Stevison, *Polym. Mater. Sci. Eng.*, 56, 401(1987).
[47] C. Carraher, A. Li, J. Kloss, A. Lombardo, *Polym. Mater. Sci. Eng*, 70, 38 (1993).
[48] C. Carraher, J. Kloss, F. Medina, A. Taylor, *Polym. Mater. Sci. Eng*, 68, 253 (1993).
[49] C. Carraher, unpublished results.
[50] F. Keene, D. Salmon, T. J. Meyer, *J. Amer. Chem. Soc.*, 98(7), 1884 (1976).
[51] C. Carraher, Q. Zhang, *Metal-Containing Polymeric Materials*, Plenum, NY, 1996, 109.
[52] C. Carraher, A. Taylor-Murphy, *Polym. Mater. Sci. Eng.*, 76, 409 (1997).
[53] C. Carraher, A. Taylor-Murphy, *Polym. Mater. Sci. Eng.*, 86, 291 (2002).
[54] C. Carraher, Q. Zhang, *Polym. Mater. Sci. Eng.*, 73, 398 (1995).
[55] C. Carraher, Q. Zhang, *Polym. Mater. Sci. Eng.*, 71, 505 (1994).
[56] C. Carraher, *Macromolecules*, 2, 306 (1969).
[57] C. Carraher, *J. Polym. Sci. Pt.* A-1, 8, 3051 (1970).
[58] C. Carraher, L. Wang, *Makromol. Chemie*, 160, 251 (1972).
[59] C. Carraher, J. Greene, *Makromol. Chemie*, 130, 177 (1969).
[60] C. Carraher, J. Greene, *Makromol. Chemie*, 131, 259 (1970).
[61] E. Franklin, *J. Chem. Soc.*, 2, 263 (1849);

[62] C. Carraher, Macromolecules Containing Metal and Metal-Like Elements Vol. 4. *Group IVB Polymers*, Wiley, Hoboken, NJ, 2005.
[63] R. Wei, L. Ya, W. Jinguo, X. Qifeng, *Polymer Materials Encyclopedia* (J. Salamone, Ed.), CRC Press, Boca Raton, FL, p 4826;
[64] M. Hoch, *Applied Geochem.*, 16, 719 (2001).
[65] C. Carraher, M. Nagata, H. Stewart, S. Miao, S. Carraher, A. Gaonkar, C. Highland, F. Li, *Polym. Mater. Sci. Eng.*, 79, 52 (1998).
[66] C. Carraher, H. Stewart, S. Carraher, M. Nagata, S. Miao, *J. Polym. Mater.*, 18, 111 (2001).
[67] C. Carraher, M. Nagata, H. Stewart, S. Miao, C. Butler, A. Gaonkar, C. Barosy, R. Duffield, F. Li, *Polym. Mater. Sci. Eng.*, 78, 36 (1998).
[68] D. Siegmann-Louda, C. Carraher, D. Chamely, A. Cardoso, D. Snedden, *Polym. Mater. Sci. Eng.,* 86, 293 (2002)
[69] I. Omae, *Organotin Chemistry*, Elsevier, Amsterdam, 1989.
[70] C. Carraher, C. Butler, Y. Naoshima, V. Forste, D. Giron, P. Mykytiuk, *Applied Bioactive Polymeric Materials*, Plenum, NY, 1990.
[71] C. Carraher, C. Butler, D. Sterling, V. Saurino, *Polym. Mater. Sci. Eng.*, 66, 352 (1992).
[72] Y. Naoshima, H. Shudo, M. Uenishi, C. Carraher, *J. Polum. Mats.*, 8, 51 (1991).
[73] C. Carraher, V. Saurino, C. Butler, D. Sterling, *Polym. Mater. Sci. Eng.*, 72, 192 (1995).
[74] C. Carraher, M. Roner, K. Shahi, Y. Ashuda, G. Barot, *JIOPM*, 18, 180 (2008).
[75] C. Carraher, T. Sabir, M. Roner, K. Shahi, R. Bleicher, J. Roehr, K. Bassett, *JIOPM,* 16, 249 (2006).
[76] D. Siegmann-Louda, C. Carraher, F. Pflueger, D. Nagy, J. Ross, *Functional Condensation Polymers*, Kluwer, NY, 2002.
[77] L. Pellerito, F. Maggio, M. Consiglio, A. Pellerito, G. Stocco, S. Grimaudo, *App. Organomet. Chem,* 9, 227 (1995).
[78] F. Maggio, A. Pellerito, L. Pellerito, S. Grimaudo, C. Mansueto, R. Vitturi, *Appl. Organomet. Chem*, 8, 71 (1994).
[79] R. Vitturi, C. Mansueto, A. Gianguzza, F. Maggio, A. Pellerito, L. Pellerito, *Appl. Organome. Chem.*, 8, 509 (1994).
[80] H. Baratne Jankovics, L. Nagy, F. Longo, T. Fiore, L. Pellerito, *Magyar Kemiai Foly.* 107, 392 (2001).
[81] R. Vitturi, B. Zava, M. Colomba, A. Pellerito, F. Maggio, L. Pellerito, *App. Organomet. Chem.*, 9, 561 (1995). (Contains listing of other related references.)
[82] L. Pellerito, F. Maggio, T. Fiore, A. Pellerito, *App. Organomet. Chem.*, 10, 393 (1966).
[83] A. Pellerito, T. Fiore, C. Pellerito, A. Fontana, R. Di Stefano, L. Pellerito, M. Cambria, C. Mansueto, *J. Inorganic Biochem.*, 72, 115 (1998).
[84] C. Carraher, C. Butler, US Patent 5, 043, 463 (Issued Aug. 27, 1991).
[85] C. Carraher, C. Butler, US Patent 5,840,760 (Issued Nov. 24, 1998)
[86] C. Butler, C. Carraher, *Polym. Mater. Sci. Eng.*, 80,365 (1999).
[87] C. Carraher, C. Butler, Y. Naoshima, D. Sterling, V. Saurino, *Biotechnology and Bioactive Polymers*, Plenum, NY, 1994.
[88] C. Carraher, C. Butler, *Biotechnological Polymers*, Technomic, Lancaster, PA, 1993.
[89] C. Carraher, C. Bytler, Y. Naoashima, D. Sterling, V. Saurino, *Industrial Biological Polymers*, Technomic, Lancaster, PA, 1995, Chpt. 8.

[90] C. Carraher, C. Butler, V. Foster, B. Pandya, D. Sterling, *Polym. Mater. Sci. Eng.*, 68, 255 (1993).
[91] C. Carraher, C. Butler, Y. Naoshima, V. Foster, D. Giron, P. Mykytiuk, *Applied Bioactive Polymeric Materials*, Plenum, NY, 1988.
[92] C. Carraher, Y. Naoshima, C. Butler, V. Foster, D. Gill, M. Williams, D. Giron, P. Mykytiuk, *Polym. Mater. Sci. Eng.*, 57, 186 (1987)
[93] C. Carraher, C. Butler, L. Reckleben, *Cosmetic and Pharmaceutical Applications of Polymers*, Plenum, NY, 1991.
[94] C.. Carraher, T. Rigdway, D. Sterling, C. Butler, *Industrial Biotechnological Polymers,* Technomic, Lancaster, PA, 1995, Chpt. 13.
[95] C. Carraher, C. Butler, L. Reckleben, A. Taylor, V. Saurino, *Polymer P*, 33(2), 539 (1992).
[96] M. Roner, C. Carraher, J. Roehr, K. Bassett, *J. Polym. Mater.*, 23, 153 (2006).
[97] G. Barot, M. Roner, Y. Naoshima, K. Nagao, K. Shahi, C. Carraher, *JIOPM*, 19, 12 (2009).
[98] C. Carraher, K. Morie, *JIOPM*, 17, 127 (2007).
[99] G. Barot, K. Shahi, M. Roner, C. Carraher, *JIOPM*, 17, 595 (2007).
[100] G. Barot, K. Shahi, M. Roner, C. Carraher, *J. Polym. Mater.*, 23, 423 (2006).
[101] C. Carraher, A. Battin, K. Shahi, M. Roner, *JIOPM*, 17, 631,2007.
[102] H. J. Wanebo, M. P. Vexeridis, *Cancer*, 78, 580 (1996).
[103] A. L. Warshaw, C. Fernandez-DelCastillo, *N. Eng. J. Med.*, 326, 580 (1992).
[104] C. G. Moertel, *Clin. Gastroenterol*, 5, 777 (1976).
[105] J. A. DeCaprio, R. J. Myer, *Clin. Oncol.*, 9, 2128 (1991).
[106] M. Moore, J. Andersen, *Proceedings ASCO*, 14, Abstract Number 473 (1995).
[107] M. L. Rothenberg, H. A. Burris, *Proceedings ASCO* ,14, Abstract Number 470 (1995).
[108] Y. F. Hui, J. Reitz, *Am. J. Health-Syst. Pharm.*, 54, 162 (1997).
[109] M. Roner, K. Shahi, G. Barot, A. Battin, C. Carraher, *JIOPM*, 19, 410, 2009.
[110] K. Shahi, M. Roner, A. Battin, C. Carraher,. *Polym. Mater. Sci. Eng.*, 99, 365, 2008.
[111] Y. Naoshima, K. Nagao, A. Battin, C. Carraher, *Polym. Mater. Sci. Eng.*, 96, 405, 2007.
[112] A. Battin, C. Carraher, *Poly. Mater. Sci. Eng.*, in press.
[113] G. Doak, L. D. Freedman, *Org. Chem. Revs.*, 6 (1970) 574.
[114] C. Carraher, M. Naas, D. Giron, D. R. Cerutis, J. *Macromol. Sci.-Chem.*, A19 (1983) 1101.
[115] C. Carraher, M. Naas, *Polym P.*, 23 (1982) 158.
[116] C. Carraher, M. Naas, *Crown Ethers and Phase Transfer Catalysis in Polymer Science*, Chpt. 7, Plenum, NY. 1984.
[117] C. Carraher, *J. Polym. Mater.* 25, 35 (2008).
[118] C. Carraher, T. Langworthy, W. Moon, *Polymer P.*, 17, 1 (1976).
[119] C. Carraher, W. Moon, *Coat. Plast.*, 34, 468 (1974).
[120] D. Siegmann-Louda, C. Carraher, Macromolecules Containing Metal and Metal-Like Elements. Vol. V. *Biomedical Applications*, Wiley, Hoboken, NJ, 2004.

Reviewed by

Louis Tisinger, Senior Scientist, Perkin-Elmer Instruments
Dorothy Steling, Sterling Environmental Consulting Company

In: Polymer Research and Applications
Editors: Andrew J. Fusco and Henry W. Lewis
ISBN: 978-1-61209-029-0

Chapter 9

DEAE-DEXTRAN AND DEAE-DEXTRAN-MMA GRAFT COPOLYMER FOR NANOMEDICINE

***Yasuhiko Onishi*[1,*], *Yuki Eshita*[2] *and Masaaki Mizuno*[3]**
[1]Ryuju Science Co. Ltd., Seto 489-0842, Japan.
[2]Dept. of Infectious Diseases, Faculty of Medicine,
Oita University, Oita 879-5593, Japan.
[3]Nagoya University, Graduate School of Medicine Program
in Cell Information Medicine, Nagoya 466-8550, Japan.

Abstract

2-diethylaminoethyl(DEAE)-Dextran is still an important substance for transfection of nucleic acids into cultured mammalian cells by the reason of its safety owing to autoclave sterilization different from lipofection vectors. However, DEAE-Dextran may not be superior to lipofection vectors with cytotoxic and a transfection efficiency. A stable soap-less latex of 2-diethylaminoethyl(DEAE)-Dextran-methyl methacrylate(MMA) graft copolymer (DDMC) of a high transfection activity has been developed as Non-viral gene delivery vectors possible to autoclave at 121 for 15 minutes. Transfection activity determined by the X-gal staining method show a higher value of 50 times or more for DDMC samples than for the starting DEAE-dextran hydrochloride and a low cytotoxic is observed for DDMC. DDMC has been also observed to have a high protection facility for DNase degradation.. The resulted DDMC on grafted MMA, to form a polymer micelle of Core-Shell particle, should become a stable latex with a hydrophilic-hydrophobic micro-separated-domain. The complex by DDMC/ DNA may be formed initially on the stable spherical structure of the amphiphilic micro-separated-domain of DDMC and have a good affinity to cell membrane for the endocytosis. The stronger infrared absorption spectrum shift to a high energy direction at around $3450 cm^{-1}$ of the complexes between DNA and DDMC compared with DEAE-Dextran may mean to form more compact structures not only by a coulomb force between the phosphoric acid of DNA and the diethyl-amino-ethyl(DEAE) group of DEAE-Dextran copolymer but also by a force from multi-inter-molecule hydrogen bond. It should conclude to DNA condensation by these inter-molecular multi-forces to be possible the higher transfection efficiency. The complex should be formed following Michaelis-Menten type equation such as $complex = K_1(DNA)(DDMC)$. It

*Email address: yasu-onishi@ryujyu-science.com

should be supported that DDMC has a strong adsorbing power with DNA because of not only its cationic property but also its hydrophobic bond and hydrogen bond. The high efficacy of this graft-copolymer autoclave-sterilized for transfection can make it a valuable tool for a safety gene delivery.

Keywords: Non-Viral Gene Delivery; DEAE-dextran; Graft Copolymer;DNA Condensation; transfection

1. Introduction

Recently, in vivo gene delivery has allowed the study of gene expression and function in animal models via insertion of foreign genes or alteration of existing genes and/or their expression patterns. However, some dangerous adverse effects such as Carcinogenic Risks remain associated with the use of viral vectors. Non-viral gene delivery vectors may be a key technology in circumventing the immunogenicity inherent in viral-mediated gene transfer. There are two types of Non-viral gene delivery carriers. one is a polyplexes composing of cationic polymer such as DEAE-Dextran, polyethyleneimine, and polybrene etc., the other is a lipoplexes composing of cationic liposomes. For the efficacy of transfection of Non-viral gene delivery, it may be very important that a complex between a carrier and DNA is formed to condense DNA protecting from DNase. Cationic polymers efficiently condense plasmid DNA into small complexes displaying a positive electrophoretic mobility. However, these cationic polymer/DNA complexes lead to cytoxity. Cationic liposomes/DNA complexes lead to only partial condensation of plasmid DNA,but have a low cytoxity . Though a polyplexes may have a large cytoxity comparison with a lipoplexes, It should be expected as a effective Non-viral gene delivery carriers because of its significantly low cytokine response in cytosol[1]. The other hand, amphiphilic polymers that bear hydrophilic-hydrophobic micro-separated-domain is also effective to include a clathrate DNA as a gene delivery carrier protected from reticuloendothelial system. DEAE-Dextran, a derivative of polysaccharide, was originally used as Non-viral gene delivery carrier by the reason of its safety owing to autoclave sterilization different from lipofection vectors. But DEAE-Dextran has a low trnsfection for large DNA and sometime cytoxity. DEAE-Dextran-MMA graft copolymer was obtained by graft-polymerizing MMA onto DEAE-Dextran in water which is very effective as a non-viral gene delivery vector[46, 47, 48, 49, 50, 51]. DEAE-Dextran-MMA graft copolymer having a hydrophilic-hydrophobic micro-separated-domain has a good affinity to a cell membrane and a high trnsfection comparison with DEAE-Dextran. DDMC also must promote the lentiviral vector activity like DEAE-Dextran[38, 39, 40, 73, 74] and protect the lentiviral vector from reticuloendothelial systems owing to its hydrophilic-hydrophobic micro-separated-domain. Moreover there are very important problem with possibility to make autoclave-sterilization for these Non-viral gene delivery vectors. On the viewpoint, it should be expected that non-viral vectors, such as the DEAE-Dextran copolymer of this chapter, are used populaly. The autoclave-sterilized DEAE-Dextran copolymer of this chapter, will increase safety by minimizing the incidence of serious diseases resulting from the immunogenicity inherent in viral vectors.

2. DEAE-Dextran

DEAE-Dextrans (diethylaminoethyl Dextrans) is a derivative of dextran reportedly been shown to reduce serum cholesterol and triglycerides. DEAE-Dextrans are obtained from synthesis of diethylaminoethyl chloride with Dextrans compose of glucose unit having mainly $\alpha - 1,6$ glycosidic linkages between glucose molecules. The DEAE-Dextran polycations prepared from Dextran 500 fractions have approximately a nitrogen content of 3 .2% which corresponds to one charged group per 3 glucosyl units. This diethylaminoethyl group which substituted on 2- hydroxy group in a glucose residue can be classified to both single and tandem groups at rate of 1 : 1.

2.1. Characteristic Properties

2.1.1. Solubility

DEAE-Dextrans are soluble in water and dilute salt solutions. Their solutions are very stable and can be sterilized by autoclaving even at low pH. Autoclaving usually can be done at $110 - 115°C$. For $20\%(w/v)$ or less, their solutions can autoclave for approximately 30 minutes; for solutions above 20%, autoclave can be done for approximately 15 minutes. But it may cool slowly.

2.1.2. Chemical Description

DEAE-Dextran is a polycationic diethylaminoethyl ether of dextran which the straight chain mainly consists of $\alpha - 1,6$ glycosidic linkages between glucose molecules, while in some cases,branches begin from $\alpha - 1,3$ linkages(and in small cases, $\alpha - 1,2$ and $\alpha - 1,4$ linkages as well). The diethylaminoethyl groups are linked to glucose residues by ether linkages. DEAE-Dextran is like a positively charged resin of DEAE-Sephadex used in ion exchange chromatography for protein purification and separation. The positively charged DEAE groups of DEAE-Dextran can bind negative ions. From the potentiometric titration curve ,there are three different acidity constant which show pKa5.5 of tertiary amino group from Tandem DEAE-DEAE group, pKa9.2 of tertiary amino group from Single DEAE group, and ~pKa 14 of tertiary amino group from Tandem DEAE-DEAE group. It is enough for DEAE-Dextran to have buffering at pH5-7 for proton-sponge effect in an endosome.

2.2. Biochemical Specification

DEAE-Dextran shows several important effects on cellular systems even at its low concentration, because its positively charged DEAE groups of DEAE-Dextran react with negatively charged cell surface. DEAE-Dextran is also used for transfecting mammalian cells with foreign DNA. When it is added to solution containing DNA meditated transfection, it binds and interacts with negatively charged DNA molecules and via Endocytosis mechanisms brings about the uptake of nucleic acids by the cell. This procedure is highly suited for transient transfection used for various gene therapy studies.

2.2.1. Enhancement of Protein and Nucleic Acid Uptake

DEAE-Dextran has been known to be effective on interaction between cells and viral nucleic acids. It is the most important effects on cellular systems that DEAE-Dextran enhances the uptake of proteins and nucleic acids.

Enhancement of protein uptake Ryser has shown that the uptake of albumin by sarcoma S-180 cells is stimulated by DEAE-Dextran[7]. Increased uptake rates for ferritin and other basic proteins have also been reported[8]. The rate of incorporation of homologous and heterologous hypoxanthine-guanine phosphoribosyltransferase into mutant chinese hamster ovary and lung cells was discussed by more than tenfold by pretreatment of the cells with DEAE-Dextran[9].

Enhancement of nucleic acid uptake With the nucleic acids, nucleotides form complexes with DEAE-Dextran by the coulomb force on the basis of their charge differences, thereby facilitating their entry into the cells[10].DEAE-Dextran enhances the uptake of nucleic acids into cells by its positively charged DEAE groups interacting with both the nucleic acid and the cell surface which negatively charged. The transport of DNA into cultured cells can be increased 3-10-fold by the use of this product[11, 12].Constantin has reported that cultured rat fibroblasts incorporated tritiated DNA in the presence of DEAE-Dextran in the G1 phase of the cell cycle[40].

2.2.2. The Enhancement of Viral Infectivity

The enhancements of viral infectivity in cell culture systems by DEAE-Dextran are well reported for a number of viruses[13, 14, 15].The incorporation of $150 - 200\mu g$ of DEAE-Dextran per ml of the agar solution, increases the sensitivity of viral plaque assays[16, 17]

2.2.3. Inhibition of Tumor Growth

It has been shown that DEAE-Dextran is capable of inhibiting the growth of various tumors in vivo if the tumor cells are incubated in a solution of DEAE-Dextran in vitro before inoculation. This inhibitory effect was found to be reversible by subsequent incubation with heparin, because it was canceled only by this incubation for five minutes after DEAE-Dextran treatment[18, 19]. Since carcinoma cells can carry a higher negative charge on the surface than normal fibroblasts, it has been postulated that the capability of DEAE-Dextran positively charged to inhibit the growth of tumors is due to electric charge alterations produced in the cell membrane structures[18, 19, 20, 21].

2.2.4. Enhancement of Interferon Production

A synthetic short double-stranded RNA such as Polyinosinic-polycytidylic acid (poly(I)-poly(C)) is widely used as a inducer in interferon production. The presence of DEAE-Dextran in the reaction medium increases the uptake of poly(I)-poly(C) by the cells and also makes the polynucleotide less susceptible to degradation from RNase. In addition, the

DEAE-Dextran/poly(I)-poly(C) complex allows to use even a lower concentration of the polynucleotide while still obtaining a given yield of interferon[22, 23, 24, 25, 26, 27].

2.2.5. Use of DEAE-Dextran as an Adjuvant

DEAE-Dextran is well established as an efficient adjuvant in vaccine production[29]. For instance, it is reported that the use of DEAE-Dextran as adjuvant for immunization of guinea pigs and swine with inactivated foot-and-mouth disease (FMD) virus, and rhesus monkeys with inactivated Venezulean equine encephalomyelitis (VEE) virus[28]. Though the mechanism of action of DEAE-Dextran is not clear, the effect on the humorous response of rhesus monkeys to inactivated VEE vaccine results in a typical IgM-IgG response. It has been speculated that DEAE-Dextran should cause a stimulation of the helper T-cell to induce its function in antibody synthesis[28, 29, 30]. On the other hand, vaccination with short-term culture filtrate proteins (ST-CFP)/DEAE-dextran induced high levels of interleukin-2 (IL-2) but low levels of interferon-gamma (IFN-gamma) from whole-blood cultures stimulated with M. tuberculosis ST-CFP in comparison with the strong IFN-gamma and IL-2 responses induced after vaccination with bacille Calmette-Guerin (BCG)[77]. It is reported that DEAE-dextran, although not as effective as Freund incomplete adjuvant(FIA), afforded some protection for the mycobacterial RNA[78].

3. The Complex between DNA and DEAE-Dextran

DEAE-dextran may be a important substance for mediating transfection of human macrophages. These transfection are accomplished via to form the complexes between DNA and DEAE-dextran. The complexes between DNA and DEAE-dextran were synthesized at various conditions changing of reaction time and pH, and evaluated by IR absorption spectra and DSC analysis. From the relation between yield(g) and pH, with reaction time 0hr,2hr,and 4hr, the optimum reaction may exist at range from pH6 to pH9. By the relation between weight ratio of DNA/DEAE-dextran in the complex and pH with the reaction time 0hr, 2hr,and 4hr, it is also shown for these ratio to be 1/1 at range from pH6 to pH9. By IR absorption spectra and DSC analysis of complexes, it was confirmed the complexes between DNA and DEAE-dextran were formed by the coulomb force between a single and tandem DEAE-group in DEAE-dextran and phosphoric acid in DNA.

3.1. The Reaction between DNA and DEAE-Dextran

DEAE-dextran was added to the DNA solution by titration with the buret and reacted. When its order reversed, the precipitaion was rarely observed. The reaction between DNA and DEAE-dextran were carried out changing from a low pH to a high pH, and the dropping time. That is, it was adjusted so that the dropping time became four hours, two hours, and 0 hours and continued reacting while stirring it. Figure 1 shows the result of a measurement of the infrared absorption spectrum of the caused precipitation. The absorption of C_2H_5 group that originates in the DEAE radical of the DEAE-dextran is near by 1500 cm^{-1}. The absorption of O-P-O group by the reverse-symmetry vibration that originates in the phosphoric acid ester of DNA was shown near by at 1230 cm^{-1}. It is also observed by about one to shift the absorption. Especially, the absorption (arrow in Figure 1) of O-P-O group by the reverse-symmetry vibration that originates in phosphoric acid ester of DNA is observed to shift to the high wave-number side along with four hours, two hours, and 0 hours of the dropping time. It is thought that an electrostatic bond strongly changes into the becoming it higher-order structure as the dropping time extends. Therefore, it is thought by this influence that the complex's it becoming close crystal condition. In addition, DNA (A) used, DEAE-dextran (C), and the complex (B) that caused, it were shown in Figure 2. It is shown near by $1500cm^{-1}$ in Figure 2 for the absorption of C_2H_5 group owing to the DEAE radical of the DEAE-dextran in the complex (B). The absorption of O-P-O group by the reverse-symmetry vibration owing to phosphoric acid ester of DNA in the complex (B) is also neighbor $1230cm^{-1}$. The absorption of O-P-O group by the reverse-symmetry vibration owing to phosphoric acid ester of DNA in the complex (B) is caused more strongly than DNA (A in Figure 2) single purpose. It is suggested that the bond caused strong uniting by electrostatic force between DEAE radical of the dextran part and a phosphoric acid in DNA part of the complex (B).

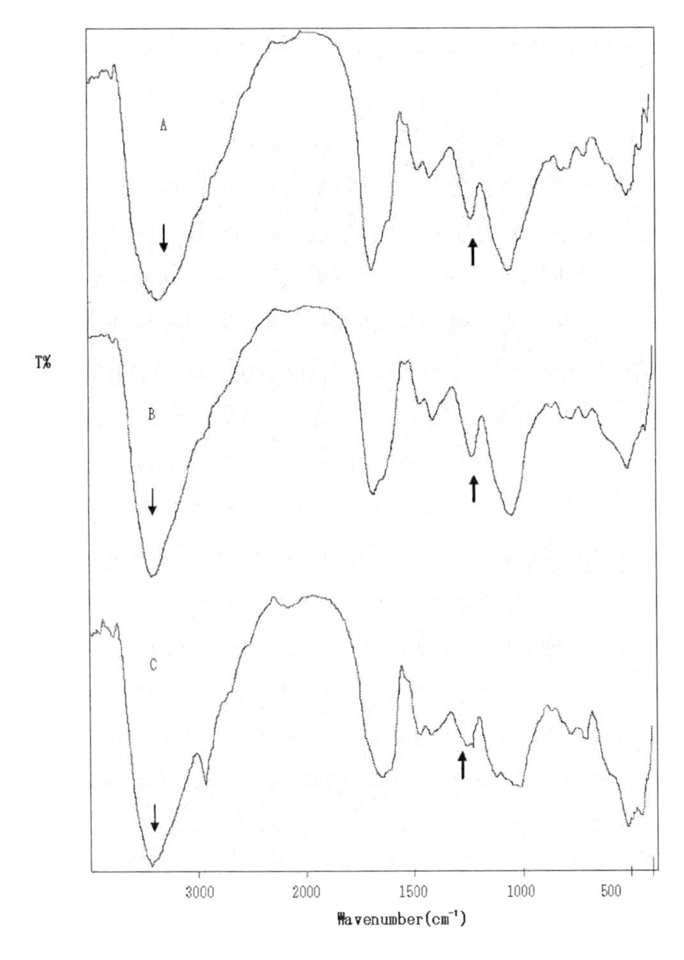

Figure 1. IR absorption spectra of complexes between DNA and DEAE-dextran at pH6: a, reaction time 0hr; b, reaction time 2hr; c, reaction time 4hr.

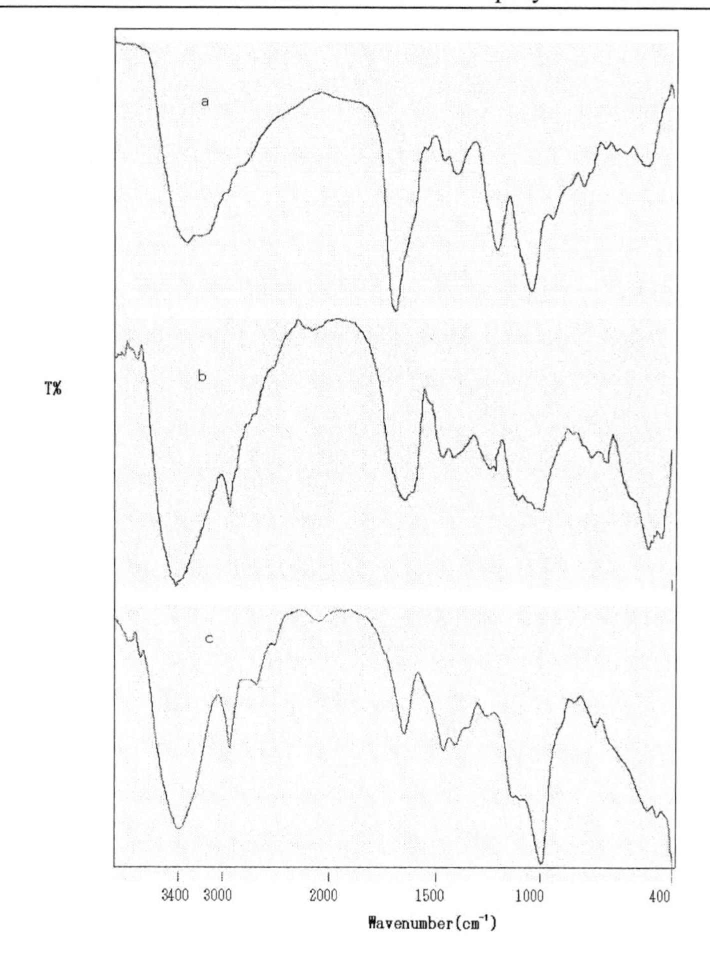

Figure 2. IR absorption spectra of DNA, the complex between DNA and DEAE-dextran, and DEAE-dextran: A, DNA;B, the complex(at pH6, reaction time 2hr); C, DEAE-dextran.

3.2. The Relation between pH Change and the Amount of Complex by DNA and DEAE-Dextran

Figure 3 shows a relation between the pH change and the amount of the complex in the dropping time of four hours, two hours, 0 hours. An interesting result was obtained respectively. Namely, the amount of complex could be observed having the maximum values of amount from pH7 in the neutral region to pH10. The dissociation of the phosphoric acid in the part of DNA is suppressed in a low pH of an acid region as for this, and the escape proton making of the DEAE radical of the DEAE-dextran proceeds in a high pH of an alkaline region. Therefore, the maximum value of the reaction is thought to be observed from pH7 in the neutral region to pH10. It is understood it has three different respective pKa from the titration curve of the DEAE-dextran. That is, pKa that originates in the tertiary amino radical that originates in single DEAE radical and tandem DEAE-DEAE radical occupies 9.2 and 5.5, respectively. The pKa that originates in the quaternary ammonium radical of tandem DEAE-DEAE radical is occupied in ‘14. This thing means that the complex can be generated in a high alkaline region, because a proton is made with tandem DEAE-DEAE radical even in a high alkaline region . That is, the precipitation generation of the complex is observed in the high alkaline region as for the pH change at the dropping time of four hours, two hours and 0 hours as Figure 3. On the other hand, the dissociation of the phosphoric acid in the part of DNA is suppressed in a low pH of an acid region. Therefore, the precipitation generation was observed at once in Figure 3 when the dropping time is 0 hours. When the DEAE-dextran was made to react at the titration time of four hours and two hours, the caused precipitation dissolved again and the precipitation was not seen though the precipitation generated. However, from this thing, the formation of the complex may be made by treble helical structure on double helical structure of DNA to which directly DEAE-dextran clathrated[35]. Moreover, it is imagined that the complex dissolves again in the state of excessive of polycation at that time.

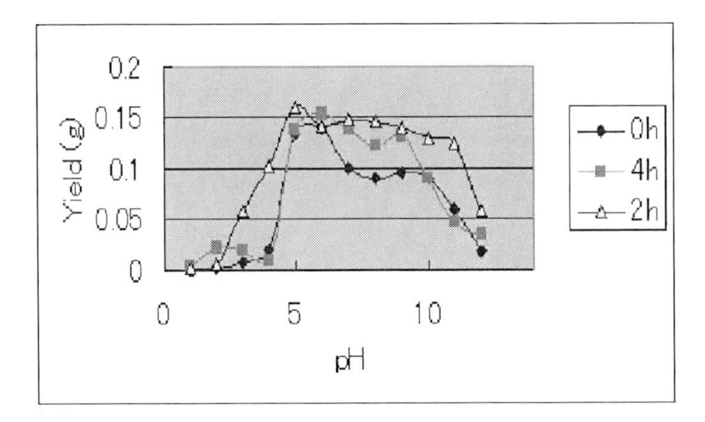

Figure 3. The relation between yield(g) and pH: reaction time 0hr; reaction time 2hr;reaction time 4hr.

3.3. The Relations between pH Change and the DNA/DEAE-Dextran Ratio in a Complex

The relations with the DNA/DEAE-dextran ratio in a complex by a pH change are shown for the dropping time of four hours, two hours and 0 hours, in Figure 4. Being common in these three dropping conditions, it is to be approximately 1 ratio for the DNA/DEAE-dextran ratio in a complex, except a low pH domain and a high pH domain. This means that this reaction goes at high reaction rate. In other words, it can be understood that reaction completely advances with all most. In addition, it is clear that the ratio depends on pH regardless of dropping time. The relations with the ratio of the electric charge ratio (N/P) of DNA/DEAE-dextran by a pH change in a complex are shown in Figure 5 for dropping time of four hours, two hours and 0 hour. According to this, the relations by a pH change with the ratio of the electric charge ratio (N/P) of DNA/DEAE-dextran of the whole complex were understood that the ratio N/P was approximately 1.28 same as the relations by a pH change with the DNA/DEAE-dextran ratio in a complex except a low pH domain and high pH domain. It is understand that this reaction goes at high reaction rate by this, and reaction completely advances with all most. In addition, it is clear that the N/P ratio of the electric charge ratio depends on pH regardless of dropping time. The reaction must advance rather than these electro-statically, and it is mainly clear to be reaction with the phosphoric acid part of the DNA by the tertiary amino group of the tandem DEAE -DEAE basis and the single DEAE basis caused by DEAE-dextran. In the low pH domain equal to pH5 or less, it is not produced a deposition almost other than dropping time for 0 hours. In addition, the DNA is about around 0.5 part to1.0 part of DEAE-dextran as shown in Figure 3. The DNA forms a complicated globule in the low pH domain equal to pH5 or less. According to the weak combination such as the hydrogen bond, it seems not combination by the reaction that is electrostatic in the low pH domain in Figure 3. However, it is thought that the complex is a poly-ion complex(PIC) between DEAE-dextran as the polycation and DNA as polyanion by the electrostatic combination as Figure 6, when thinking about the measurement the result of the above-mentioned infrared absorption spectrum at any place other than the low pH domain.

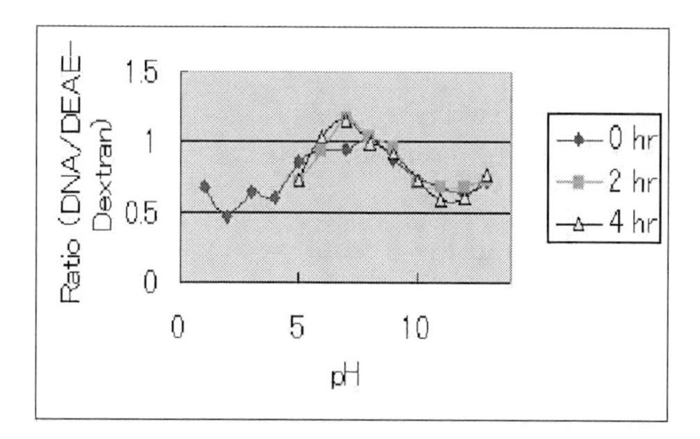

Figure 4. The relation between the weight ratio of DNA/DEAE-dextran in the complex and pH: reaction time 0hr; reaction time 2hr; reaction time 4hr.

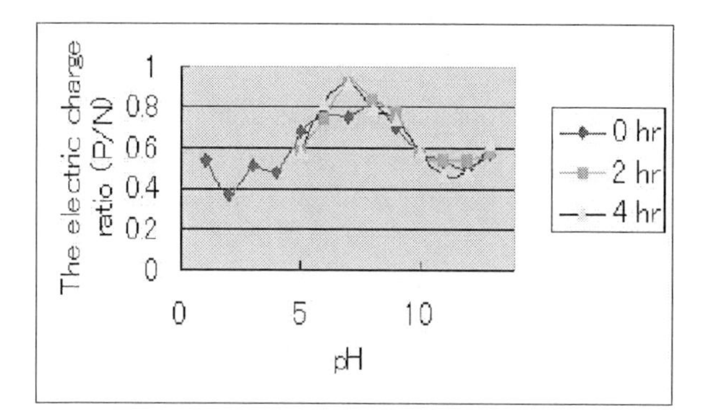

Figure 5. The relation between the electric charge ratio of DNA/DEAE-dextran in the complex and pH: reaction time 0hr; reaction time 2hr; reaction time 4hr.

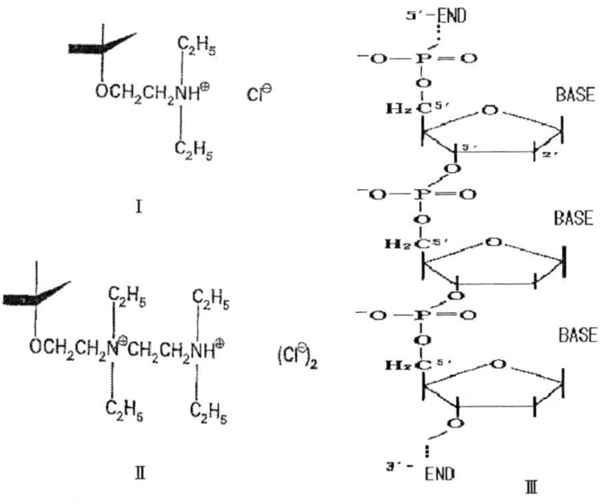

Figure 6. Shematic representation of the reaction between DNA and DEAE-dextran:I, single DEAE-group; II , tandem DEAE-group; III , DNA.

3.4. The Thermal Analysis of the Complexes by DNA/DEAE-Dextran (DSC Analysis)

DSC of the complex by DEAE-dextran/DNA which generated by dropping time of four hours, two hours, and 0 hours, in a condition of pH 6 each, were analyzed. The results are shown in Table1. There are a broad endothermic peak which is thought to be an adhesion water and an adsorption water each. As dropping time gets longer about the complex which generated for four hours, two hours, and 0 hours, its endothermic amount ΔH become small as $8.3mJ/mg$, $1.1mJ/mg$, and $0.6mJ/mg$. This develops their triple helical structure as a result of DEAE-dextran entwining each other with the DNA, and the structure of the complex changes for highly ordered structure as dropping time gets longer, and it seems that it can not keep adhesion water and adsorption water by taking the structure of the compact cohesion state. The reaction sediment is observed only in the case of 0 hours of dropping time at the low pH domain, but dropping time does not contradict this fact because of this structure composed by a hydrogen bond. Their deposition which produced by titration time for four hours and two hours dissolves again and are not seen. In other words it is imagined that the complex forms the dense crystal state as dropping time becomes long (as reaction time becomes long). There are shown the measurement results of the infrared absorption spectrum of the complex which generated in the condition of pH 6 for four hours, two hours, and 0 hours of dropping time corresponding to Figure 1. The absorption by the reverse symmetry stretching vibration of P - O caused by a phosphoric acid ester part of the DNA is watched in the vicinity of $1,230cm^{-1}$. The absorption of the C_2H_5 group coming from the DEAE radical or the DEAE-DEAE radical is shown at the neighborhood of $1,500cm^{-1}$. The absorption by the stretching vibration of - NH group coming from the DEAE radical or the DEAE-DEAE radical is watched in the vicinity of $3,450cm^{-1}$. All their absorption shifts to the high energy side are observed according to their longer dropping time. It is thought that electrostatic combination become strong with extension of dropping time from this thing. It is thought that the complex becomes a dense crystal state with this influence.

Table 1. Results of DSC analysis for the complex between DNA and DEAE-dextran at pH6.0

Reaction time(hr)	0	2	4
Absorption H(mJ/mg)	8.3	1.1	0.6
Onset temperature()	82.0	85.0	107.1

4. The Complex between RNA and DEAE-Dextran

DEAE-dextran is an important substance used for transfer of nucleic acids into cultured mammalian cells. The transfection of a mammalian cell is accomplished via the complexation between nucleic acids and DEAE-dextran. Such complexes were thus synthesized using the RNA from a yeast under various conditions including reaction time and pH, and analyzed by IR absorption spectroscopy and DSC. From the relation between the yield, pH, and reaction time, the optimum pH was in the range 5-10. The relation between the weight ratio of RNA/DEAE-dextran in the complex, pH, and reaction time indicated the ratio to be 0.25-0.55 in the pH range 5-10. The IR absorption spectra and DSC curves confirmed that the complexes were formed by the coulomb force between a single and tandem

4.1. The Formation Reaction of Complex by DEAE-Dextran and RNA

As well as the case of DNA, the DEAE-dextran solution is dropped to the RNA solution with the buret. It is DEAE-dextran in the RNA solution because it doesn't cause precipitation easily if reverse in order. They were made to drop the DEAE-dextran solution with the buret and to react. Changing the pH from a low pH to a high pH and the dropping time, the complex by the DEAE-dextran/RNA was formed. The dropping time of DEAE-dextran solution to RNA solution has been changed. That is, it was adjusted that the dropping time became four hours, two hours, and 0 hour and continued the complex formation reaction while stirring it. It is just straight dropping DEAE-dextran solution to RNA solution from the buret to say 0 hours in the dropping time here. The caused precipitation was observed for four hours, two hours, and 0 hour of dropping time and became the more close hard when taking the more longer dropping time . Figure 7 shows the result of a measurement of the infrared absorption spectrum of the caused precipitation. The absorption of C_2H_5 group that originates in the DEAE radical or the DEAE-DEAE radical of the DEAE-dextran is shown near by $1500cm^{-1}$. The absorption based on their pyranose ring is shown in the vicinity of $1000cm^{-1}$'$1100cm^{-1}$. The absorption of O-P-O group by the reverse-symmetry vibration originated in the phosphoric acid ester of RNA is shown around 1230 cm^{-1}. The absorption by the stretching vibration of - NH group coming from the DEAE radical or the DEAE-DEAE radical is watched in the vicinity of $3,450cm^{-1}$. As for the absorption around 1230cm^{-1}(upper arrow in figure) by the reversely symmetric stretching vibration of O-P-O due to phosphoric acid ester of the RNA and the absorption nearby 3 ,400cm^{-1} (arrow in figure in the under) by stretching vibration of N-H group coming from DEAE radical or the DEAE-DEAE radical of DEAE-dextran, what shift to the high energy side according to longer dropping time of four hours, two hours, and 0 hour, are observed. It is thought with electrostatic combination to be changing into the high level structure which becomes strong and dense following the extension of the dropping time like a case of the DNA. This was confirmed by visual observation during an experiment concretely, but, as for a complex being in a dense condition with dropping time, it is thought with this reason. Furthermore, RNA (a), DEAE-dextran (c) and the complex (b) which resulted are shown in Figure 8. In the complex (b), there can be observed the characteristic absorptions of DEAE-dextran and the RNA.The absorption of C_2H_5 group coming from DEAE group of DEAE-dextran is the vicinity of 1 ,500cm^{-1}.The absorption based on their pyranose ring of DEAE-dextran is shown in the vicinity of 1000 cm^{-1}'1100cm^{-1}. The absorption by the reverse symmetry expansion and contraction vibration of O-P-O caused by phosphoric acid ester of the RNA is observed in the vicinity of 1 ,230cm^{-1}. Here, as the absorption by the reverse symmetric expansion and contraction vibration of O-P-O due to phosphoric acid ester is not particularly strong different from the case of DNA, the electrostatic combination of the RNA complex is estimated not like a case of the DNA.

4.2. The Relations between pH Change and the Complex Yield by RNA and DEAE-Dextran

As for Figure 9, the relations between pH change and the complex yield are shown at dropping time of four hours, two hours, and 0 hour. An interested result was provided each. From pH 5 to pH 10 of the neutral level, the maximum of the yield was observed . As for this, the dissociation of the polyphosphoric acid of the RNA controlled in the low pH of the acidity area and the deprotonation of the DEAE group of DEAE-dextran advanced in the high pH of the alkalinity area, the maximum of the reaction can be observed in the neutrality area from pH 5 to pH 10,but reaction seems to be accelerated rapidly when pH reaches pH 5 of the neutrality area different from the case of DNA. DEAE-dextran was understood to have three different pKa from a neutralization titration curve each. In other words their pKa due to the tertiary amino radical due to single DEAE group shows 9.2, and pKa caused by the tertiary amino radical of the tandem DEAE-DEAE group shows 5.5, and pKa caused by the quaternary ammonium group shows in '14. This means that the reaction by RNA and DEAE-dextran are strongly influenced by the tertiary amino radical of the tandem DEAE-DEAE group. In addition, the tandem DEAE -DEAE group of DEAE-dextran shows that composite generation is possible like a case of the DNA even to the high alkalinity area because of their protonating to a high alkaline level. With the relations between pH change and the complex yield for dripping time of four hours, two hours, and 0 hours, the composite deposition generated is observed to the high alkalinity area in Figure 9. the other hand, dissociating it for the polyphosphoric acid moiety of the RNA was controlled in the low pH of the acid level, and it followed that it was hard to produce deposition. When the complex yield compares it in the case of DNA, there is a little it generally. The formation of the RNA complex seems to be different from mechanism of

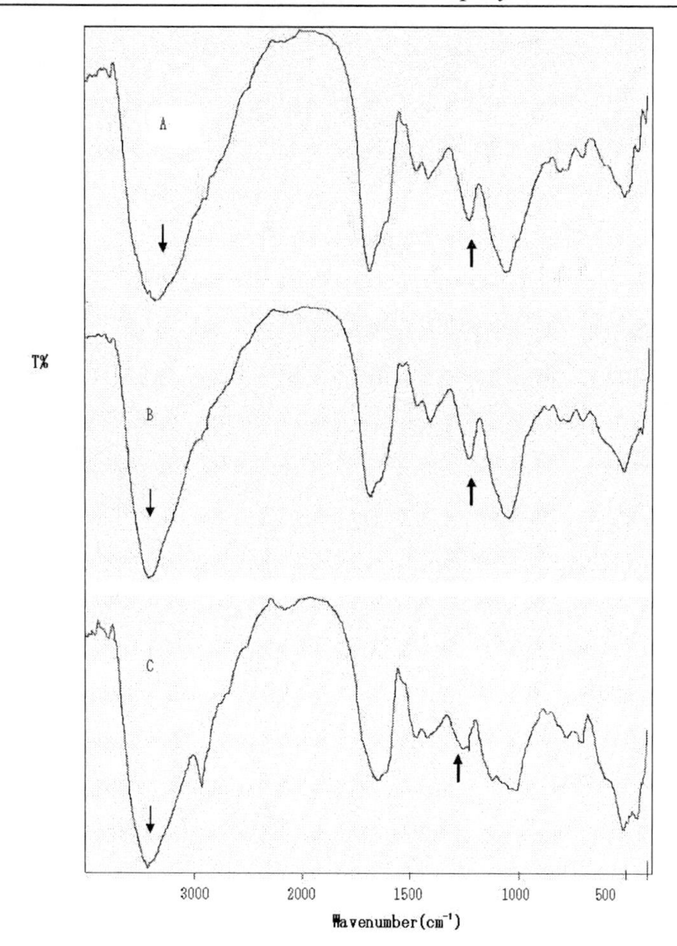

Figure 7. IR absorption spectra of complexes between RNA and DEAE-dextran at pH6: A, reaction time 0 h; B, reaction time 2 h; C, reaction time 4 h.

a clathrate directly, and forming helical structure such as the case of the DNA rather than these[35]. In the case of the low molecular RNA, it will be made an inclusion so that the RNA which extended does stacking in DEAE-dextran different from the case of DNA.

4.3. The Relations between pH Change and the RNA/DEAE-Dextran Electric Charge Ratio (N/P) and the Relations between pH Change and the RNA/DEAE-dextran Weight Ratio in the Complex

The relations between pH change and the RNA/DEAE-dextran weight ratio in the complex are shown in Figure 10 for the dripping time of four hours, two hours, and 0 hour. In the case of the DNA except a low pH domain and a high pH domain approximately one weight ratioFIt was 1. The weight ratio that DEAE-dextran of the RNA faced in three dropping conditions was generally low with around 0.25‘ 0.55. This means that this complex reaction is not worse at such high reaction rate. In addition, the weight ratio seems to become almost around 0.3 constantly if dropping time becomes long. The relations between

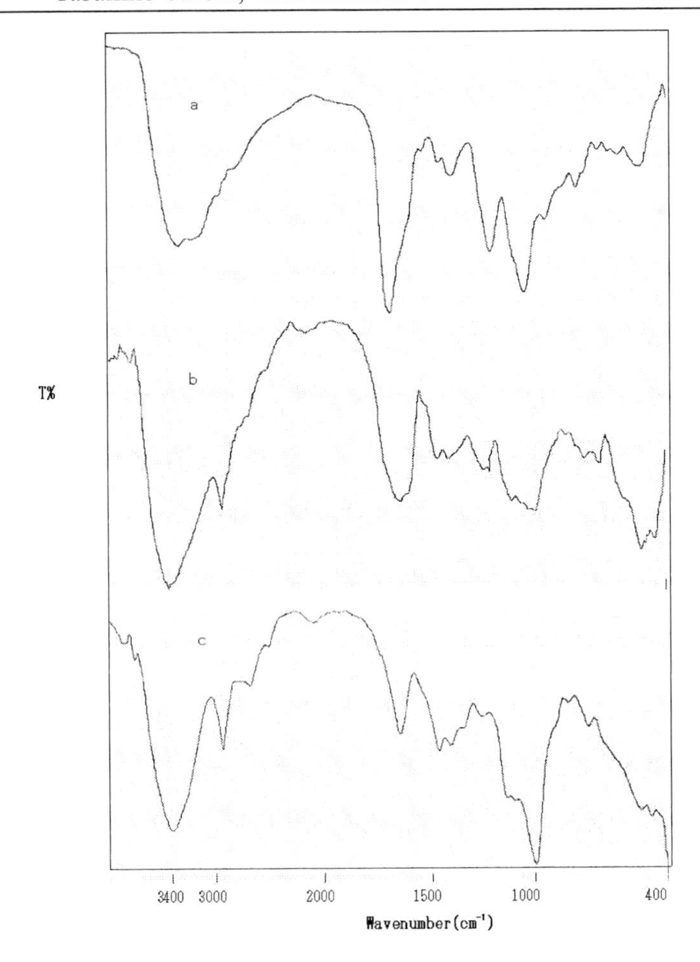

Figure 8. IR absorption spectra of RNA, the complex between RNA and DEAE-dextran, and DEAE-dextran: a, RNA;b, the complex(at pH6, reaction time 2 h); c, DEAE-dextran

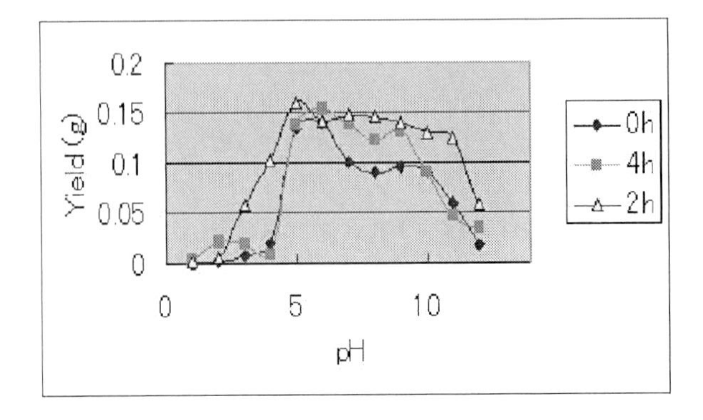

Figure 9. The relation between yield(g) and pH with the reaction of RNA/DEAE-dextran: reaction time 0 h; reaction time 2 h;reaction time 4 h

pH change and the electric charge ratio (N/P) of RNA/DEAE-dextran in the complex are shown in Figure 11 for the dripping time of four hours, two hours, and 0 hours. According to this, the relations between pH change and the electric charge ratio (N/P) of RNA/DEAE-dextran in the complex cannot have any tendency and has the value of the electric charge ratio (N/P) that seems to be around 3.2 maximums in two hours and 2.7 maximums in 0 hours of dripping time, The other hand, it seems to become almost uniformity with around 2 to 2.7 in four hours of dripping time. Thinking about their electric charge ratio (N/P) of approximately 1.28 in the case of the DNA except a low pH domain and a high pH domain, their an anionic group and a cationic group must react at a constant ratio by their electron charge. Because an anionic group and a cationic group do not correspond with their electron charge for this reaction, it is understood that this reaction is worse in quite low yield. Additionally, there are many 2'-hydroxyl groups to the RNA different from DNA in RNA, and its molecular weight is low. From these facts, it will disturb composite deposition reaction that hydration characteristics are big. DEAE-dextran oneself which is polycation approximately completely settle in the good domain of the yield from pH 5 to pH 10 when regarding relations with the RNA/DEAE-dextran electric charge ratio (N/P) and the RNA/DEAE-dextran weight ratio. It is sure that the reaction advances rather than these electro-statically, and it must be hydrogen bond or a hydrophobic bond to promote their other inter-molecules cohesion. The reaction here must be mainly caused by electrostatic combination between the tertiary amino radical of single DEAE group and tandem DEAE - DEAE group of DEAE-dextran and the phosphoric acid part of the RNA. In addition, there are many 2'-hydroxyl groups (to the second place carbon of the ribose) that the RNA is different from DNA, and molecular weight is low, so that the reaction do not almost produce deposition as in Figure 9 because the hydration characteristics are big to be apparent. According to the weak combination such as the hydrogen bond, like the said article in the low pH domain equal to or less than pH 5, but there will be a few benefits in stead of combination by electrostatic reaction in the low pH domain. However, it can be considered that after all the complex compose of DEAE-dextran as the polycation and RNA as the polyanion to be a poly-ion complex(PIC) by mainly their electrostatic combination when thinking about the results of a measurement of the above-mentioned infrared absorption spectrum any place other than the low pH domain.

4.4. The Thermal Analysis of the Complexes by RNA/DEAE-Dextran (DSC Analysis)

The complexes by RNA and DEAE-dextran forming in dropping time of four hours, two hours, and 0 hours at the condition of pH 6 were analyzed by differential scan calorimeter (DSC). The results are shown in Table2. There are a broad endothermic peak which is thought to be an adhesion water and an adsorption water each. As dropping time gets longer about the complex which generated for four hours, two hours, and 0 hour, its endothermic amount ΔH become small as $33.5mJ/mg$, $32.9mJ/mg$, and $25.3mJ/mg$. As dropping time becomes long, this change into the high ordered structure so that the composite structure is dense and shrink each as a result of DEAE-dextran entwining each other with the RNA, and it seems that it can not keep adhesion water and adsorption water by taking the structure of the compact cohesion state. However, it is hard to think that the structure coheres strongly

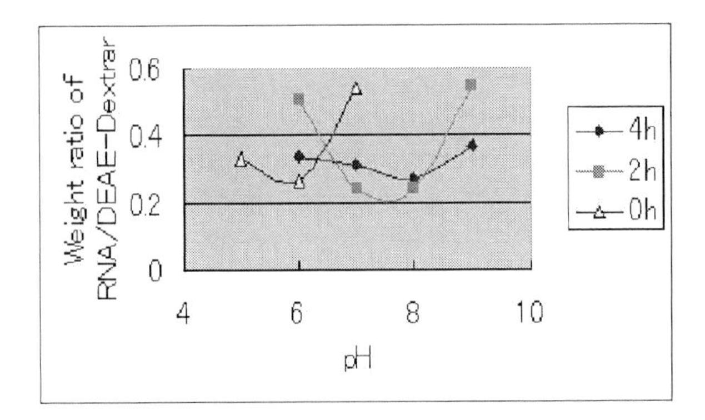

Figure 10. The relation between the weight ratio of RNA/DEAE-dextran in the complex and pH: reaction time 0 h; reaction time 2 h; reaction time 4 h.

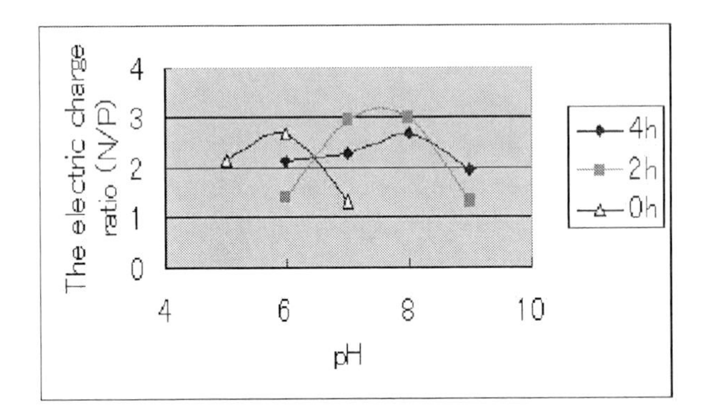

Figure 11. The relation between the electric charge ratio of RNA/DEAE-dextran in the complex and pH: reaction time 0 h; reaction time 2 h; reaction time 4 h.

and is in a dense condition more to compare it in the case of DNA. Figure 7 shows the result of a measurement of the complex composite by infrared absorption spectrum that generate corresponding to the dripping time in four hours , two hours, and 0 hour, at the condition of pH 6. The absorption of the C_2H_5 group coming from the DEAE radical or the DEAE-DEAE radical is shown at the neighborhood of $1,500^{-1}$. The absorption by the reverse symmetry stretching vibration of P-O caused by a phosphoric acid ester part of the RNA is observed in the vicinity of $1,230cm^{-1}$. The absorption by the stretching vibration of - NH group coming from the DEAE radical or the DEAE-DEAE radical is observed in the vicinity of $3,450cm^{-1}$. All their absorption shifts to the high energy side are observed according to their longer dropping time. It is thought that electrostatic combination become strong with extension of dropping time from this thing. It is thought that the complex becomes a dense crystal state with this influence. It was small when the degree of the shift compared it in the case of DNA, but the combination that was electrostatic from this thing became strong with extension at dripping time, and it was the same tendency that a complex was in a dense condition even if the degree was low. Among gene delivery-systems, as for this composite of the complex dense state, it will become an important factor strongly influenced when it is penetrated its cell membrane by endocytosis and exhausted from the transportation endoplasmic reticula to the cytoplasm in the cell or it is taken in the nucleus by nuclear film penetrated. Finally, it seems that their transportation efficiency strongly depend on the conformation of complex for gene delivery-system,

Table 2. Results of DSC analysis for the complex between RNA and DEAE-dextran at pH6.0

Reaction time(hr)	0	2	4
Absorption H(mJ/mg)	33.5	32.9	25.3
Onset temperature()	39.5	45.1	50.7

5. DEAE-Dextran-MMA Graft Copolymer

A stable and soapless latex of 2-diethylaminoethyl (DEAE)-dextran-methyl methacrylate (MMA) graft copolymer (DDMC) was, for non-viral gene delivery vectors, possible autoclave at 121 for 15 minutes. Their transfection activity was measured using the X-Gal staining method and a higher value of 50 times or more was for DDMC samples as compared with the starting DEAE-dextran hydrochloride. DDMC was also confirmed that a high level of protection for DNase degradation. This led DDMC with amphiphilic domain to a polymer micelles of core-shell particles which should contribute to stable latex with a hydrophilic-hydrophobic micro-separated-domain . The complex by DDMC and DNA may be formed initially on the stable spherical structure of the amphiphilic micro-separated-domain of DDMC and have a good affinity to cell membrane for the endocytosis. The infrared absorption spectrum shift to a high-energy direction at about 3450 cm^{-1} of complexes of DNA and DDMC may mean more compact structures not only by a Coulomb force between the phosphoric acid of DNA and the diethyl-amino-ethyl (DEAE) group of DEAE-dextran copolymer but also by a force of multi-inter-molecular hydrogen bonding. They should be based on the DNA condensation to allow for the higher transfection efficiency. The high efficiency of this graft-copolymer autoclave-sterilized can make it a valuable tool for a safety gene delivery.

5.1. Theoretical

A polyion complex [ES] by polycation and DNA should be set up, followed by Michaelis-Menten type reaction:

$$[E]+[S] \overset{Km}{\rightleftharpoons} [ES] \tag{1}$$

Where [E] is polycation and [S] DNA. Then, in a steady state, we have

$$[E][S]/[ES] = k_2/k_1 = Km \tag{2}$$

where Km is a equilibrium constant. The initial concentration of the polycation is $[E_0]$, we have

$$[E] = [E_0] - [ES] \tag{3}$$

With Eq. (2) we get

$$[ES] = [E_0][S]/(Km + [S]) \tag{4}$$

For DDMC of core-shell particles, when a low concentration of [E] and [S], the kinetic k1[E][S] may be small, because a Coulomb force between the phosphoric acid from nucleic acids and the diethyl-amino - ethyl (DEAE) group of DEAE-dextran should not be large enough to compare with hydrogen bond in its PIC structure and the hydrophobic force of the graft poly (MMA) for its conformation. That is why Km is much larger than [S]. As $Km \gg [S]$, we have

$$[ES] = [E_0][S]/Km \tag{5}$$

For DEAE-dextran as a linear water soluble polymers, at a low concentration of [E] and [S], the kinetic $k_1[E][S]$ is large because it is by primarily a Coulomb force between the phosphoric acid of the nucleic acids and 2-diethyl-amino-ethyl (DEAE) group of DEAE-dextran. That is the reason why [S] is much larger than Km. As $Km \ll [S]$,the concentration of the the complex is almost initial concentration of $[E_0]$

$$[ES] = [E_0] \tag{6}$$

5.2. Materials and Methods

5.2.1. Preparation of DDMC

The samples of DDMC1, DDMC2 and DDMC3 in Table3 were as described in the following procedures: 2-diethylaminoethyl (DEAE)-dextran hydrochloride (nitrogen content 3%) from dextran with Mw500, 000 was placed in a water, and then methyl methacrylate (MMA) was added. With stirring, the air in the reaction vessel was fully replaced by N_2 gas. To the solution was added CAN and 0.1N nitric acid, and the mixture was reacted with stirring for 1 hour at 30. Then, hydroquinone was added to the reaction solution, and the resulting latex DDMC water was purified by water dialysis using a cellophane tube in order to remove the un-reacted MMA, ceric salts, and nitric acid.

5.2.2. Characterization of DDMC

The DDMC is stable and soap-less in water. However, the DDMC caused by methanol are insoluble in water and acetone at 25. As DEAE-Dextran hydrochloride is soluble in water and poly (MMA) is soluble in acetone, it is obvious that the DDMC is not a mixture but copolymer of DEAE-dextran and poly(MMA). The infrared absorption spectrum of DDMC as shown in Figure 12 has some characteristic absorption bands at 1730 cm^{-1} and at 1000cm^{-1} to 1150cm^{-1}, based on the carbonyl group of poly(MMA) and the pyranose ring DEAE-dextran , respectively. The DDMC exhibits different solubility from DEAE-dextran and poly(MMA) and shows the above-described characteristic absorption in the infrared absorption spectrum. From this fact, it is assumed that DDMC is graft-polymerized. These graft copolymers have an amphiphilic domain to form a polymer micelle by their hairy nano-particles and should become a stable latex of core shell particles with a hydrophilic-hydrophobic micro separated domain to form a spherical structure[52].

Table 3. Properties of DEAE-dextran-MMA Graft Copolymers

Sample	Weight-increase(%)[1]	Precipitation time(hrs) by DNA
DDMC1	150	2.0
DDMC2	100	1.0
DDMC3	130	1.5
DEAE-dextran	0	96.0

[1] Weight increase (%)=(weight of MMA used/weight of DEAE-dextran hydrochloride used) 100

5.2.3. Reaction between DNA and DDMC

The in vivo interaction between DNA and basic proteins such as histones, known by the occurrence of a partially unfolded part of chromatin, plays an important role in the regulation of gene-transfer system[53]. For the reaction between DNA and DDMC, to 1ml of DNA (Salmon Sperm EX) solution (20 mg / ml), 2 ml of DDMC solution (10mg/ml) was drop-wised to precipitate the complex of DDMC / DNA. The obtained complex was insoluble in water, which is a good solvent for nucleic acids. Thus, the complex between DNA and DDMC must consist of a poly-ion complex (PIC) by a Coulomb force on a polymer micelles[63]. In the case of the samples in Table3, a complex between DNA and DDMC2 with a weight increases of 100% by grafted is required 1 hour precipitation. The complex between DNA and DDMC1 with 150% weight increases by grafted required 2 hours of precipitation, respectively. However, a complex between DNA and DEAE-Dextran hydrochloride required 96 hours in order to precipitate at this condition. This precipitation facility for the polyfection will be very important to transport DNA to cytoplasm in a cell. From thermodynamics for the complex reaction between DNA and DDMC, the Gibbs energy change $\Delta G[Jg^{-1}]$ of the complex reaction is as follows:

$$\Delta G = \Delta H - T\Delta S \tag{7}$$

Where $\Delta H[Jg^{-1}]$ is the enthalpy change and $\Delta S[Jg^{-1}K^{-1}]$ is the entropy change on formation of the complex by DNA / DDMC. The Gibbs energy change $\Delta G[Jg^{-1}]$for the complex formation should be minus because of its great plus entropy change (ΔS) by a hydrophobic effect of hydrophobic domains of poly (MMA) in DDMC, because the enthalpy change $\Delta H[Jg^{-1}]$ is small in comparison with the entropy change (ΔS). As reflected in Table3, this means that the complex reaction between DNA and DDMC can go easily and the molecule of the complex in a crystal structure can be more densely packed in comparison with the complex of DEAE-dextran/DNA. This may be that the complex of DDMC / DNA is compactly formed not only by a Coulomb force between the phosphoric acid of DNA and 2-diethyl-amino-ethyl (DEAE) group of DEAE-dextran copolymer but also by a force of the multi-inter-molecular hydrogen bond and hydrophobic force from thc hydrophobic areas of the graft poly (MMA) in DDMC and it has also a high level of protection for DNase as shown in Figure 20.

5.2.4. The Structure of the Complex between the DNA and DDMC

The transition of DNA structure, namely a coil-globule transition, induced discrete on/off switching of transcriptional activity[54]. This collapse transition in individual DNA chains has been reviewed as DNA condensation[55, 56]. The in vitro collapse of DNA may be l induced by various cationic vectors such as cationic lipids[57, 58, 59], peptides[60], or cationic polymers[61, 62]. In the case of cationic lipid vectors, the complex of dioctadecylamidoglycylspermine (DOGS) / DNA, which has a nucleosome-like structure in which DNA wraps around a micellar aggregate of DOGS and has a connection with each other to this network structure ,is very effective for gene transfection[54]. The complex by cationic polymer/DNA in the cytoplasm be generally protected from restriction enzymes for the collapse of DNA[55, 56]. In the case of the cationic dextran, it is reported the complex of DEAE- Dextran-DNA in the cytoplasm can be also protected from DNase[39]. Formation of a complex of nucleic acids (DNA or RNA) and cationic graft-copolymers, such as DEAE-dextran-MMA copolymer, is represented by a Coulomb force between the phosphoric acid from nucleic acids and 2-diethyl-amino-ethyl (DEAE) group of DEAE-dextran-MMA copolymer. The so obtained complex of DDMC/DNA was insoluble in water which is a good solvent for nucleic acids. The complex by DDMC / DNA is a poly-ion complex (PIC) of a polymer micelles that why the infrared absorption spectrum of νOH vibrations vibrations and νNH at around $3100 - 3800m^{-1}$ is shown in Figure 13. These results show that the complex between DNA and DDMC must form a poly-ion complex through the Coulomb force. The complex of nucleic acids (DNA or RNA) and Cationic Graft Copolymer from this research have a not typical but special poly-ion complex[63]. Relatively DDMC a poly-ion complex between DNA and DDMC may be proceed by the hydrophobic force graft poly (MMA) as a function of its large positive entropy change (S) in comparison with the enthalpy change (H). The complex of DDMC/DNA should be thought to provide a more compact structure through the shift of νNH vibration at $3450cm^{-1}$ in comparison with the DDMC as shown in Figure 12. Then, DDMC was confirmed that a high level of protection for DNase degradation as Figure 20. As shown in Figure 22, the resulting DDMC of the hairy nano-particles[64, 65], having the amphiphilic domain to form a polymer micelles of core-shell particles [65], should become a stable latex with a hydrophilic-separated micro-domain[66, 67]. The complex of DNA and DDMC may be initially formed on the spherical structure of the amphiphilic micro-separated-domin of DDMC with a good affinity for the cell. If (e) DNA is in the cytosol, the complex structure can be protected from degradation of DNA and Dextransucrase to make it easier for (f) Nuclear entry to occur as shown in Figure 21.

5.2.5. Measurement Infrared (IR) Absorption Spectra for DDMC

IR measurements on samples KBr method using Jasco FT/IR-300. Figure 12 shows the infrared absorption spectra of the resulting complex between DNA and DDMC2. The spectrum of the complex has some characteristic absorption bands at 1730 cm^{-1}, $1220cm^{-1}$, and in 1000 to $1150cm^{-1}$, based on the carbonyl group of poly (MMA), the P-O stretching vibration , and the pyranose ring of DEAE-dextran , respectively. Figure 13 also shows the infrared spectrum of the complex DDMC2 / DNA and the complex DEAE-dextran/DNA by comparison with DDMC2 and DEAE-dextran, respectively. The spectrum of the complex has some characteristic absorption bands due to hydrogen bonds by -OH and - NH at about $3100 - 3800m^{-1}$. There was also a structural difference between the complex by DDMC/DNA and the complex by DEAE-dextran/DNA by the decrease of $\nu NH - O$ vibration at about $3350cm^{-1}$ and the shift of νNH vibration from $3450cm^{-1}$ to $3550cm^{-1}$ as in Figure 13. This shows the complex by DDMC / DNA should provide a stable and strong PIC multi-inter-molecular hydrogen bond in comparison with the complex by DEAE-dextran/DNA to reduce collapse transition in DNA chains. This may be due to not only a Coulomb force between the phosphoric acid of DNA and the diethyl-amino-ethyl (DEAE) group of DEAE-dextran copolymer but also a force of multi-inter-molecular hydrogen bonding. They should be based on the DNA condensation of DDMC possible a high transfection efficiency to reduce the collapse transition in the DNA chains[54].

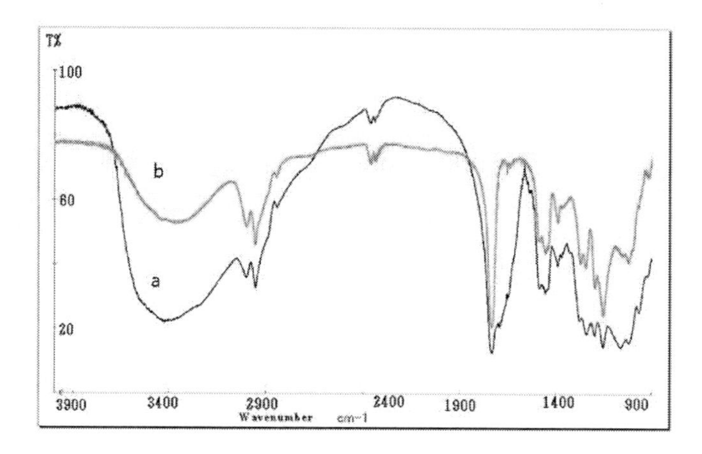

Figure 12. IR absorption spectra of DEAE-dextran-MMA Graft Copolymer and the complexes between DNA and DEAE-dextran-MMA Graft Copolymer: a, complex of DDMC2/DNA; b, DDMC2.

6. DEAE-Dextran-MMA Graft Copolymer for Nonviral Delivery Carrier

6.1. Transfection

Recently, in vivo or in vitro gene delivery, the investigation of gene expression by insertion of foreign genes or the modification of existing genes has been allowed. Some harmful adverse effects are associated with viral vectors. Non-viral gene delivery vectors should be a key technology to circumvent immunogenicities inherent in viral-mediated gene transfer. 2-Diethylaminoethyl (DEAE)-dextran was used for a non-viral gene delivery vector by the reasons for its fast and simple procedures and security for autoclave sterilization different from other lipofection vectors[36, 37]. It can be also better for the gene therapy that the DEAE-dextran transfections offer only temporary. However, DEAE-dextran may not be superior to viral vectors and lipofection vectors with cytotoxic and a transfection efficiency.

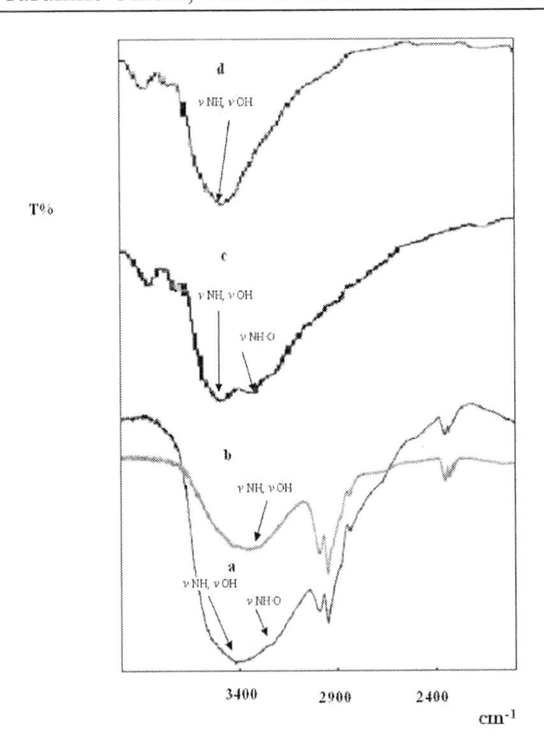

Figure 13. IR absorption spectra: a, complex of DDMC2/DNA; b, DDMC2; c, complex of DEAE-dextran/DNA; d, DEAE-dextran.

Many efforts have been done in the area of non-viral gene delivery vector, for high transfection efficiency and safety, especially with DEAE-dextran[38, 39, 40]. DEAE-dextran has been investigated for its transfection efficiency, and some good conditions for human macrophages has been found[41]. DEAE-dextran, as a strong adsorbing properties with DNA or RNA due to its cationic properties, is found to change its force for adsorbing nucleic acid by pH and ionic strength[42, 43]. It is also known that dextran-MMA graft copolymer with a hydrophilic-separated micro-domain has a good affinity to a cell membrane[44]. This case is associated with new graft copolymers, the number of possibilities for a non- viral gene delivery carriers, from a DEAE-dextran and a vinyl monomer. DEAE-dextran-methyl methacrylate (MMA) graft copolymer (DDMC) by graft-polymerizing MMA on DEAE-dextran in water using ceric ammonium nitrate(CAN) is a stable latex[45], which proved to be very effective as a non - viral gene delivery vector [46, 47, 48, 49, 50, 51].

6.1.1. Cell Line and Cell Culture

We have HEK293cell line which is a permanent line of primary human embryonic kidney by sheared human adenovirus type 5 DNA[45]. The cell line was grown in Dulbecco's Modified Eagle Medium (DMEM), supplemented with 10% fetal calf serum, 0.1 mM non-amino acids, 5mM L-glutamine and antibiotics (streptomycin, $100 \mu g/ml$, penicillin, 100U/ml). We also COS7 which is a SV40-transformed African Green Monkey kidney fibroblast cell line from the cells of CV-1 (simian) origin. SV40 can produce Large T antigen, but has a bug in the genome replication. The cell lines were in Dulbecco's modified Eagle's medium-Glutamax supplemented with 10% fetal calf serum.

6.1.2. Plasmid DNA

A pCAGGS / LacZ, which expresses -galactosidase in eukaryotic cells, was inserted CAG promoter plasmid, pCAGGS. The plasmids were amplified in Escherichia coli DH5 and purified by Qiagn Mega plasmid purification kit (Qiagen). Another plasmid DNA, pGL3-Control vector (Promega) with the Luciferase gene and SV40 promoter/enhancer elements was used.

6.1.3. Transfection of DDMC/DNA

Procedure For the transfection of DDMC/DNA, HEK293cells (150×10^4 cells) were seeded in to 35-mm culture dishes and incubated at 37 in a humidified atmosphere of $5\%CO_2$ the day before the transfection. In a sterile tube, a $10\mu g$ DNA was diluted in 270μl 1 PBS. A $14\mu l$ of autoclaved DDMC with a concentration of 10mg/ml except DDMC3(for DDMC3, as 10mg/ml DEAE-dextran) was added to the DNA solution. Then, briefly mixed by vortexing. The growth medium was removed from cells which are transfected. The cells were washed twice with 1 PBS. DDMC/DNA solution was added to the cells to swirl the dish to distribute. The dish was careened slowly several times to ensure complete coverage of the cells, and incubated at 37 for 30 minutes. The dish was slowly careened several times during the incubation. A growth medium of 1 ml was added and incubated at 37 for 48-72 hours. After incubation, the measurement of transfection activity was carried out using the X-Gal staining method. Following transfection protocol, the transfection of HEK293 by sample DDMC1, DDMC2 and DDMC3 were carried out with plasmid DNA.

Thixotropy Property An aqueous solution of DEAE-dextran graft copolymer is with a thixotropy property as a non-Newtonian fluids, whose transfection solution is sufficient to flow and wet cells through a sheer stress for the solution in the transfection, such as swirling the pan distribute. A water-soluble polysaccharides and copolymers of vinyl connection are interested for the wet-ability of the cells by the rheological properties[69]. Most reports in the rheological field were on the water-soluble polysaccharides and a few report deals with Graft polymerization of acrylamide (AM) on water-insoluble polysaccharides such as cellulose[70].

6.1.4. Protocol

Protocol A(Seed COS cells 16 to 20 hours earlier at 8×10^5 per 100 mm diameter dish)

1. Prepare cells by plating the day before the transfection .
2. Prepare the wash solution (1 PBS(phosphate-buffered saline)). Warm wash solution and the cationic graft-copolymer to 37 .

3. Using the 10 PBS prepared, dilute to a 1 solution. Prepare transfection solutions as outlined:100mm plate: In a sterile tube, dilute 20μg of DNA (plasmid corded Luciferase activity) to 540μl in 1 PBS. Add 28μl of the cationic graft-copolymer having the 10mg/ml as the starting polycation to the DNA solution#. Tap the tube to mix.

4. Remove culture medium* from the cells. Wash cells twice with 2 10ml per 100mm plate..

5. Add the mixture between DNA and the cationic graft-copolymer to cells. Swirl plate to distribute**.

6. Incubate plates at 37 for 30 minutes with occasional rocking**.

7. Gently add 6ml of growth medium* per 100mm plate. Incubate for up to 2.5 hours or until cytotoxicity is apparent***. Change medium. Cells are generally ready to harvest 48-72 hours post- transfection and assayed for luciferase activity.

8. The luciferase activity was determined by a Luciferase assays kit. For example, using a kit of Promega(Promega,Madison,WI) and a Turner model TD-20e luminometer, the luciferase activity was reported in Turner light units(TLU). Cells were lysed in the culture plate wells with 200μl of lysis buffer per well and the cell lysates transfered to microfuge tubes. The cell lysates were centrifuged to pellet insoluble cellular debris and 20μl aliquots of the cell lysates were assayed in 100μl of luciferase activity. The approximation of a Turner light units(TLU) was done by assaying serial dilutions of recombinant luciferase (cat. # E170A,Promega,Madison,WI) as recommended.

*Dulbecco's modified Eagle's medium supplemented with 10% fetal calf serum, 0.1 mM nonessential amino acids, 5 mM l-glutamine, and antibiotics (100$\mu g/mL$ streptomycin, 100 U/mL penicillin) **An aq. solution of cationic graft-copolymer having a thixotropy property, a strong sheer stress is needed for its solution to flow and wet the cell. It is better to use Incubator shaker. ***If necessary, 0.1μl-5μl of SDS supplement for toxi-blocking can be added to medium until cytotoxicity is apparent. # Do not reverse its order.

Protocol B(Seed HEK293 cells 16 to 20 hours earlier at 3×10^5 per 35 mm diameter dish)

1. Prepare cells by plating the day before the transfection .

2. Prepare the wash solution (not supplied; either 1 PBS or 1 HBSS) Warm wash solution and the cationic graft-copolymer 37 .

3. Using the 10 PBS prepared, dilute to a 1 solution. Prepare transfection solutions as outlined: Per 35mm plate: In a sterile tube, dilute 10μg of DNA to 270μl in 1 PBS#. Add 14μl of the cationic graft-copolymer having the 10mg/ml as the starting polycation to the DNA solution. Tap the tube to mix.

4. Remove culture medium* from the cells. Wash cells twice with 2 2.0ml per 35mm plate.

5. Add the DNA/ (the cationic graft-copolymer) mixture to cells. Swirl plate to distribute**.

6. Incubate plates at 37 for 30 minutes with occasional rocking**.

7. Gently add 3.0ml of growth medium* per 35mm plate. Incubate for up to 2.5 hours or until cytotoxicity is apparent***. Change medium. Cells are generally ready to harvest 48-72 hours post transfection and assayed for transfection activity.

*Dulbecco's modified Eagle's medium supplemented with 10% fetal calf serum, 0.1 mM nonessential amino acids, 5 mMl-glutamine, and antibiotics (100 μg/mL streptomycin, 100 U/mL penicillin) **An aq. solution of cationic graft-copolymer having a thixotropy property, a strong sheer stress is needed for its solution to flow and wet the cell. It is better to use Incubator shaker. ***If necessary, 0.1μl-5μl of SDS supplement for toxi-blocking can be added to medium until cytotoxicity is apparent. #Do not reverse its order.

6.1.5. Charge Ratio of DNA/DDMC

Charge ratio of between DNA (PF5.33%) and DDMC3 (NF1.4%,)can be calculated as mol ratio of P/N as below:

$$P/N = (y \times 0.0533 \times 14)/(x \times 0.014 \times 31) \tag{8}$$

Where / is weight ratio of DNA/ DDMC and atomic weight is P=31 and N= 14. Table4 shows the charge ratio of between DNA and DDMC3 with transfection of HEK293 cells by DDMC3 with grafting rate of 130% for sample 1 of 10mg/ml, sample 2 of 20 mg/ml, and the sample 3 of 28.5mg/ml. As the charge ratio of DNA/DEAE-dextran (10mg/ml) for the transfection of HEK293 cells is 0.013, the transfection of the sample 2 and sample 3 may be reasonable compared with an initial DEAE-dextran.

Table 4. Charge ratio of DNA/DDMC at the transfection of HEK293 cells

	Charge ratio of DNA/DDMC (P/N)
DEAE-dextran 10.0 mg/ ml	0.013
DDMC3(grafting rate 130)	
sample 1 10.0 mg/ ml	0.030
sample 2 20.0 mg/ ml	0.015
sample 3 28.6 mg/ ml	0.011

6.1.6. Transfection Efficiency

As shown in Figure 14, the transfection efficiency, the transfection activity was measured using the X-Gal staining and a higher value was for the samples of DDMC1 and DDMC2 than for the starting DEAE-dextran hydrochloride. In Figure 15, levels of -galactosidase enzyme are 50-times and more -gal/protein [mU ml/mg] for the case of DDMC3 (sample1,

Sample2, Sample3) compared with DEAE-dextran. From the results, the transfection efficiency and the reaction of the formation of the complex increase more on the use of DDMC hydrochloride instead of DEAE-dextran hydrochloride.

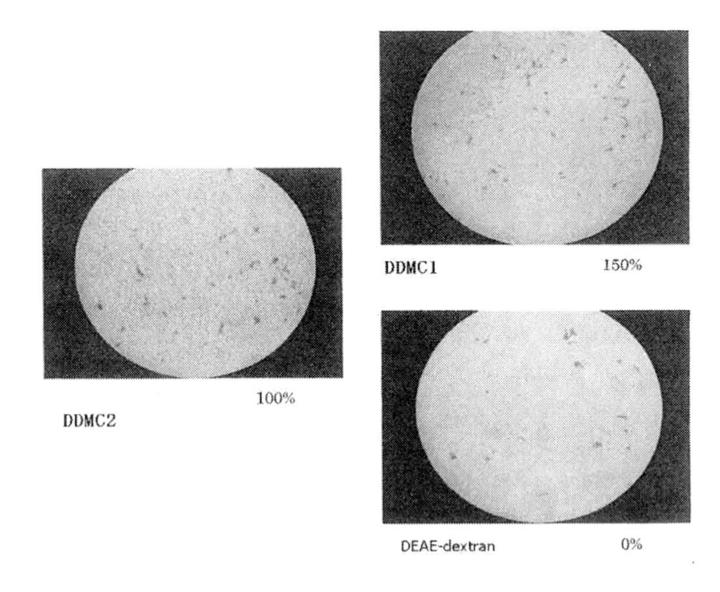

Figure 14. Transfection of a monolayer of HEK 293 cells by DEAE-dextran-MMA Graft Copolymer: DDMC1, weight increase 150%; DDMC2, weight increase 100%; DEAE-dextran, weight increase 0%.

6.1.7. Cytotoxicity for the Transfection

Figure 15 shows levels of -galactosidase enzyme from transfection of HEK293 cells by DDMC3 with grafting rate of 130% for sample 1 of 10mg/ml, sample 2 of 20 mg/ml, and the sample 3 of 28.5mg/ml. The quantities of -galactosidase enzyme increases with the concentration of DDMC3 as 31.13, 32.27 ,and 57.24 [mU ml/mg] for sample1, sample2, sample3, respectively. Figure 16 also shows the transfection of COS7 cells by DEAE- dextran and DDMC3 with grafting of 130% for sample 1 of 10mg/ml, sample 2 of 20 mg/ml, and the sample 3 of 28.6mg/ml. The transfection of COS7 cells was by pGL3DNA/DDMC. Cells are ready to harvest 72 hours after transfection and assayed for luciferase activity. Their luciferase activity increase with increasing concentrations of DDMC3 as $sample1 < sample2 < sample3$. These show a low cytotoxicity for DDMC3. Figure 17 shows the transfection activity of HEK293 cells with the X-Gal staining of the sample 1 of 10mg/ml, sample 2 of 20mg/ml, and the sample 3 of 28.6mg/ml of DDMC3 compared with DEAE-dextran, same as Figure 15. Figure 17 also shows that many cell deaths by apoptosis are in HEK293 cells for DEAE-dextran. However, Figure 17 shows little cell death by apoptosis or necrosis in HEK293 cells for the sample 1, sample 2 and sample 3 of DDMC3. These show a low cytotoxicity for DDMC3 compared with DEAE-dextran. Figure 18 shows the change in the efficiency of transfection with 2-fold volume, for example using $20\mu g$ of DNA, of both the DNA and DDMC as far as the protocol. The transfection of HEK293 cells by sample DDMC1 and DDMC2, with 2-fold amount of DNA and DDMC as far as the protocol, has shown 2-times higher efficiency than an original from a transfection activity with the help of X-Gal - staining method. From these results, as impossible for DEAE-dextran, a cytotoxicity of transfection can be confirmed to reduce and improve if DDMC used instead of DEAE-dextran hydrochloride.

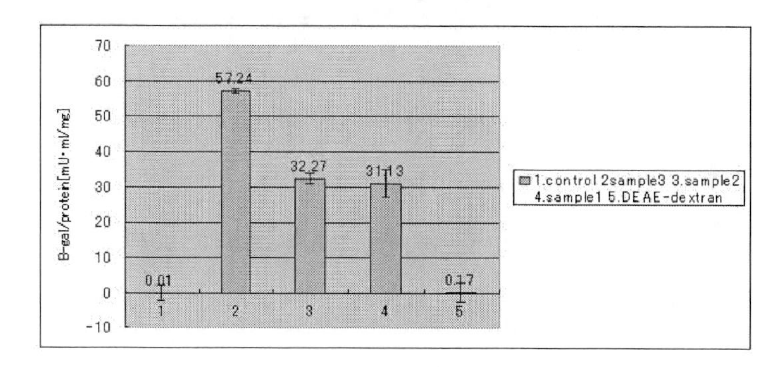

Figure 15. -Galactosidase Assay: DDMC3(sample1 10.0mg/ml, sample2 20.0mg/ml, sample3 28.6mg/ml), weight increase 130%; DEAE-dextran(10mg/ml), weight increase 0%. Each condition represents the mean of three transfecrions per condition. Standard deviation was calculated and plotted for each condition.

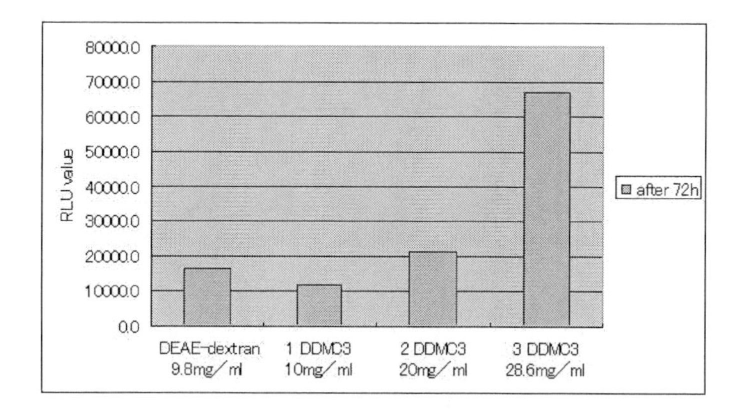

Figure 16. Transfection of COS7 cells by sample of DEAE-dextran and DEAE-dextran-MMA graft copolymer having grafting rate of 130% for sample 1 of 10mg/ml, sample 2 of 20mg/ml,and sample 3 of 28.6mg/ml. The transfection for COS7 cells was carried out by pGL3DNA/DDMC.Cells are ready to harvest 72 hours post- transfection and assayed for luciferase activity.

6.1.8. Kinetics for the Transfection

The complex between DNA and DDMC should be followed by Eq. (9) of Michaelis-Menten equation such as Eq. (5), because it is not only by a Coulomb force between the phosphoric

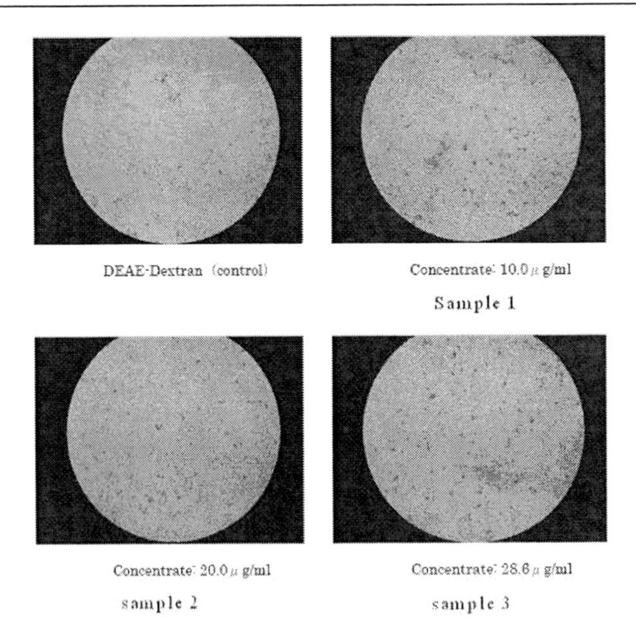

Figure 17. Transfection of a monolayer of HEK 293 cells by DEAE-dextran-MMA Graft Copolymer(DDMC3) of weight increase 130% for sample 1 of 10mg/ml, sample 2 of 20mg/ml,and sample 3 of 28.6mg/ml.; DEAE-dextran, weight increase 0% , for sample of 10mg/ml .

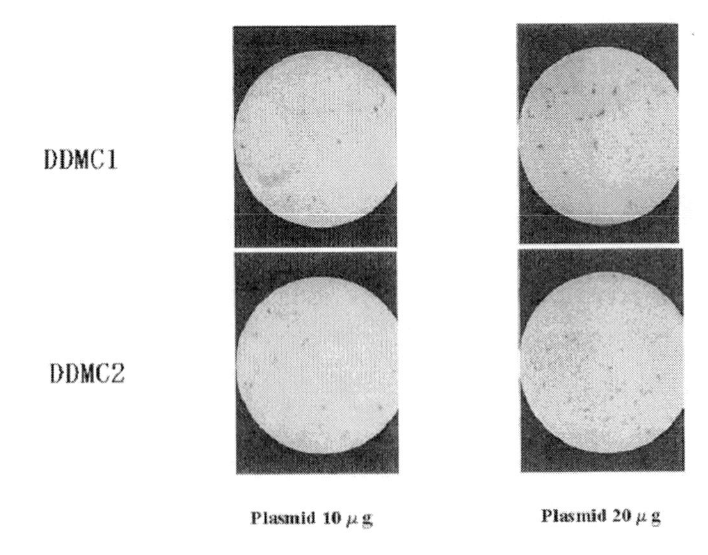

Figure 18. Effect of cytotoxicity by DEAE-dextran-MMA Graft Copolymer on transfection of a monolayer of HEK 293 cells: DDMC1, weight increase 150%; DDMC2, weight increase 100%. Figure 18 shows the change of transfection efficiency of HEK293 when using 2 times quantity, for example using 20μg DNA and 28μl of the autoclaved DDMC having a concentration of 10mg/ml, of both DNA and DDMC as much as the protocol.

acid from nucleic acids and the diethyl-amino-ethyl (DEAE) group of DEAE-dextran, but also by hydrogen bond in its structure and the hydrophobic bond of the Graft Poly(MMA).

$$complex = Km_1(DNA)(DDMC) \tag{9}$$

The complex between DEAE-dextran and DNA should be followed by Eq. (10) of the first order in the concentration of DEAE-dextran as Eq. (6), because it is by primarily a Coulomb force between the phosphoric acid from nucleic acids and the diethyl-amino-ethyl (DEAE) group of DEAE-dextran.

$$complex = Km_2(DEAE - dextran) \tag{10}$$

Figure 19 shows the activity of the transfection of COS7 cells by samples of DEAE-dextran and DDMC3 with grafting of 130% for DNA $0.075\mu g$ when the highest relative luciferase expression (RLU) in each experiment was set to 100%. We have calculated the RLU values of Eq. (9) and Eq. (10) in Figure 19 with $Km_1 = 1055 \times 10^{-7}(g/well)^2$ and $Km_2 = 1626 \times 10^{-5}\mu g$ obtained at maximum RLU value, respectively. As a quantum of the complex, of course, in proportion to RLU value, Eq. (9) and Eq. (10) can be correctly estimated.

6.1.9. Protection against DNase Degradation

The complex of DDMC/DNA is thought to be a compact structure by its shift of νNH vibration at $3450 cm^{-1}$, as shown in Figure 12. DDMC was also considered to have a high level of protection for DNase degradation as Figure 20. DNase I degrades both double- and single-stranded DNA endonucleolytically, with production of 3'-OH oligonucleotides.

Measurement of DNase Degradation Toluidine blue (TB) and DNA form a complex of stains reaction[67]. The stain reaction between TB and DNA was adjusted by adding 1 ml of 0005% TB solution (pH 7.0) to 1 ml DNA solution (10 mg/ml salmon sperm EX, Wako, Osaka, Japan). For the reaction of DDMC/DNA, to 2 ml DNA solution mixed with TB solution, the solution of DDMC (10 mg/ml DEAE-dextran) was added to reduce the complex of DDMC / DNA stained by TB. The resulting precipitate of the complex of DDMC/DNA stained by TB was separated through filters (ADVANTEC 5A, Toyo Roshi, Tokyo, Japan), then in 4 ml of distilled water, mixed with 0.01 ml (10 units) of DNase (RQ1 RNase-free DNase, Promega, Madison, WI) and 0.1 ml of 10 Reaction Buffer (400 mM Tris-HCl, $100 mM MgSO_4$, $10 mM CaCl_2$, pH 8.0) , and incubated at 37 C for 6600 minutes. DNase degradation was determined by measuring the absorbance for TB isolation of DNA in the water with a spectrophotometer. TB in water is broadly absorbed in the red zone, with a maximum absorbance at 633 nm and a shoulder near 600 nm The wavelength at which the absorption was followed as the 633 nm was enough to a sufficiently large initial absorption.

DNase Degradation TB is isolated in water from the DNA on the degradation of the DNA is stained by TB. Figure 20 shows the absorption for TB isolated from the DNA of the samples in water with a spectrophotometer. Figure 20 also shows the rate of DNase degradation with a slope of the absorption and the time, since Beer law plot between the

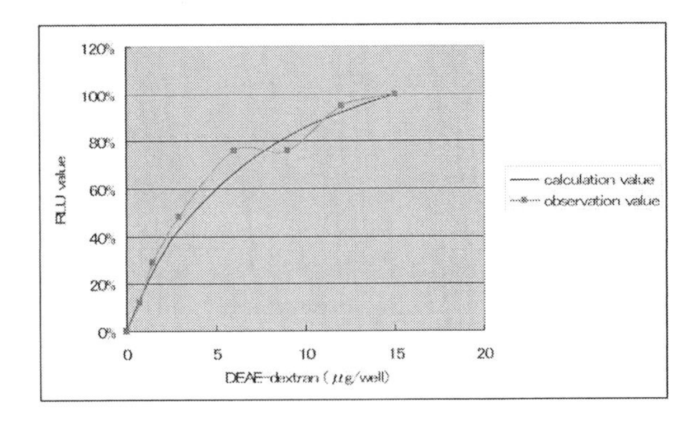

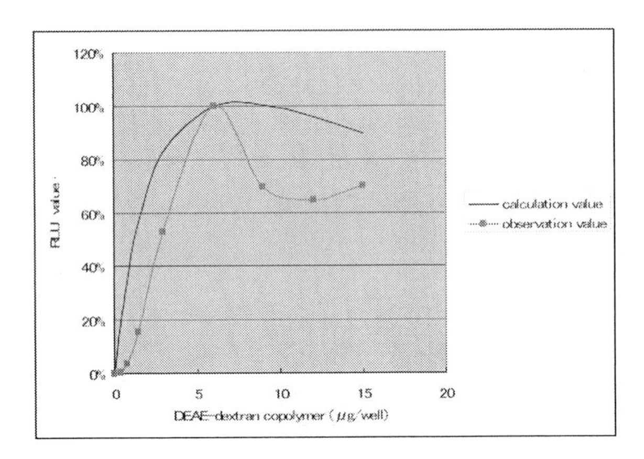

Figure 19. Transfection of COS7 cells by samples of DEAE-dextran and DEAE-dextran-MMA graft copolymer (DDMC3) having grafting rate of 130% for DNA 0 .075μg. Maximum luciferase expression within each experiment was set to 100%.

absorbance and the concentration of TB can be applied[72]. In Figure 20, some degradation of DNase may occur for the sample DEAE-dextran/DNA, but the DDMC3/DNA sample shows a very low DNase degradation. We have also found that the DNA in the colloidal phase of interaction with DEAE-dextran was protected against DNase degradation[39, 73, 74]. As a result, DDMC apparently has a higher protective effect on DNase than DEAE-dextran.

6.1.10. DNA Delivery Path Ways to Cytoplasm or Nucleus by DDMC

As shown in Figure 21, DNA delivery path ways to cytoplasm or nucleus by DDMC for the transfection of cells should be carried out using the following steps: (a) the formation of a complex between DNA and DDMC. (b) Uptake. (c) Endosytosis (endosomes). (d) Escape from endosytic vesicles. (e) Release of DNA in the cytosol. (f) Nuclear entry. (g) DNA release and transmission in the nucleus. For the transfection efficiency, it is very important to examine how Uptake in step (b), the nuclease resistance in step (c), Escape from endosytic vesicle in step (d), Nuclear entry in step (f), and DNA released in step (g). The

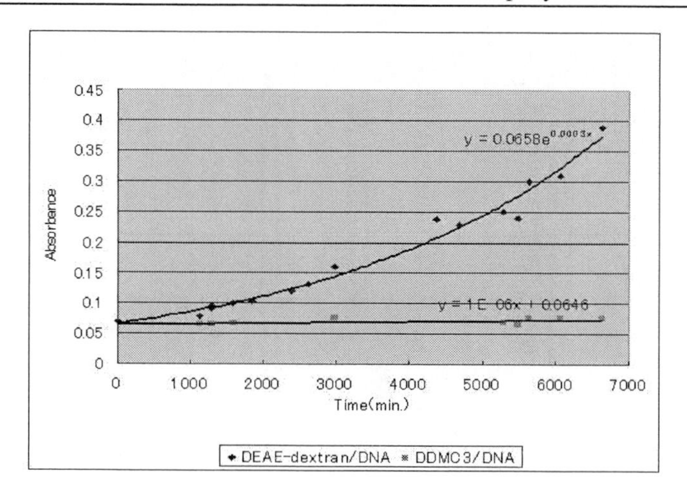

Figure 20. DNase degradation:The samples were added 4ml of a distilled water, then 10μ of DNaseT(RQ1 RNase-Free DNase, Promega)and 0.1ml of 10 Reaction Buffer(400mMTris-HCl, 100mM $MgSO_4$, 10$mMCaCl_2$, pH8) and incubated at 37.The wavelength used for this experiment was 633nm for Toluidine Blue isolated from DNA.

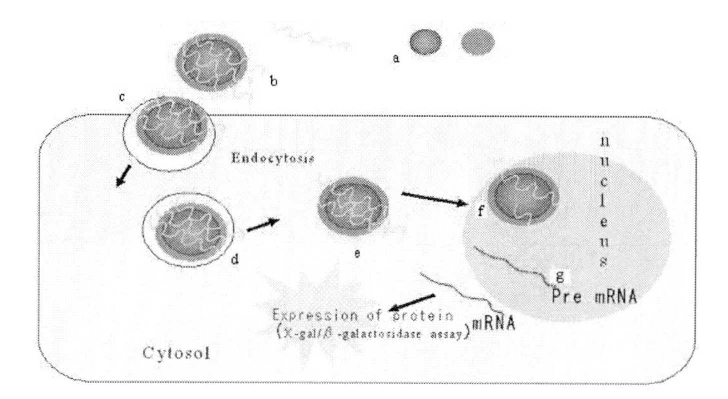

Figure 21. Schematic drawing of DNA delivery path ways by DDMC

positively charged DEAE-dextran copolymer firstly interacts with the negatively charged phosphate backbone of DNA. The resulting complex in step (a) is absorbed by endocytosis in the cells. The specific molecular structure of DDMC with a positive charge and a hydrophilic-hydrophobic micro-separated area ensures easy entry of DNA into cell for the steps (b), (c), (d), (f) and (g), and in particular DNA penetration of the plasma membrane and intracellular release of DNA in these steps because of its affinity to a cell membrane by the hydrophilic-hydrophobic micro-separated-domain[44]. DDMC with a thixotropy property as a non-Newtonian fluids, whose transfection solution should effectively help to flow and wet cells through a sheer stress for overcoming the barrier of transfection, as the steps (c). With respect to the transfection efficiency for all steps was also considered important to the protection of DNA degradation by DNase I attack. However, DNA is tightly packed in native genomes through the formation of the complex in step (a) and the way the

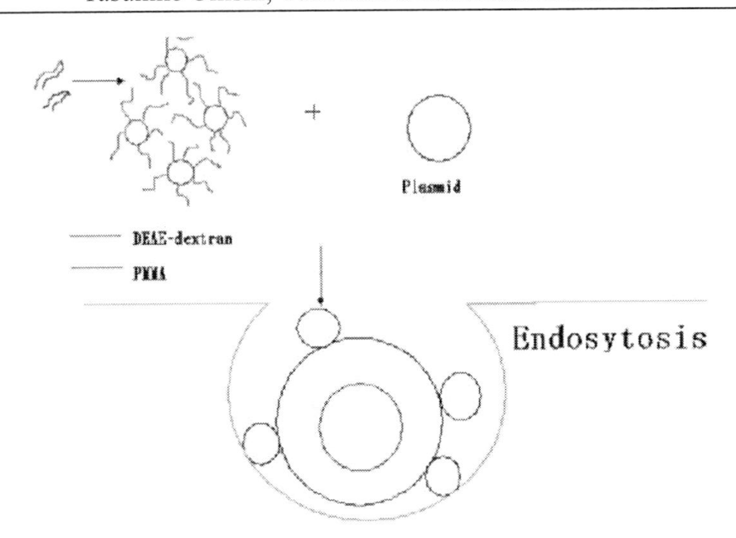

Figure 22. Schematic representation of endosytosis by the complex between DDMC and Plasmid. The complex of DDMC /plasmid may be initially formed on the spherical structure of the amphiphilic micro-separated-domain of DDMC by a Core-Shell type polymer micelle

packaging is involved in the mechanism of gene expression. The formation of the complex should induce the structural transition of DNA, what called DNA condensation by coil-globule transition[54, 56, 75, 76]. This may be discrete ON/OFF switch in transcriptional activity[56] and can be a protection against the degradation of entrapped DNA and associated complexes, as the result of Figure 20

7. Conclusions

In a first step (a) the delivery path, such as the formation of a complex of nucleic acids (DNA or RNA) and DDMC is, of course, with a Coulomb force between the phosphoric acid from nucleic acids and the diethyl-amino-ethyl (DEAE) group of DEAE -dextran. Figure 12 also shows the infrared absorption spectra of the resulting complex between DDMC (sample DDMC2) and DNA. The spectrum of the complex has some characteristic absorption bands at $1730cm^{-1}$, $1220cm^{-1}$, 1000 to $1150cm^{-1}$, and $3450cm^{-1}$ corresponding to the carbonyl group of poly (MMA), PO vibrations in the DNA, the pyranose ring of DEAE-dextran in DDMC and diethyl-amino-ethyl (DEAE) group of DEAE-dextran in DDMC, respectively. The absorption spectrum shift at around $3450\,cm^{-1}$ of the complexes should lead to more compact structures by mainly a Coulomb force between the phosphoric acid of DNA and the diethyl-amino-ethyl (DEAE) group of DEAE-dextran in DDMC to conclude DNA condensation theory[56]. In addition, there are also many differences between the complex of DDMC/DNA and the complex of DEAE-dextran/DNA with the hydrogen bond in its structure by the decrease of $\nu NH - O$ vibration at about $3350\,cm^{-1}$ and the shift of vibrations νNH from $3450cm^{-1}$ to $3550cm^{-1}$ in Figure 13. As shown in Figure 22, which the DDMC led its hairy nanoparticles with amphiphilic domain to a polymer micelles of core-shell particles, it should contribute to stable latex with a hydrophilic-hydrophobic micro-separated-domain[63]. The complex by DDMC and DNA plasmid may be formed initially on the spherical structure of the amphiphilic micro-separated-domain of DDMC to be stable and have a good affinity to cell membrane. If (e) Release of DNA in the cytosol, this structure should be protected from degradation by DNase and dextransucrase so as to be easy for the (f) Nuclear entry as shown in Figure 21. From these viewpoints, DDMC has a lower cytotoxicity and higher transfection efficiency compared with DEAE-dextran. The higher transfection efficiency of this graft-copolymer autoclaved can make it a valuable tool for gene delivery in vivo or in vitro.

8. Concluding Remark

Specifications of DDMC Vector from DEAE-Dextran MMA Copolymer were as below:

1. Stable for autoclaving Sterilization (FDA 121, 15min.)
2. Fast and easy procedure
3. Applicable in high-throughput-screening (HTS)
4. No serum inhibition
5. High efficiency by use of small DNA amounts
6. An excellent reproducibility
7. A very low toxicity in comparison with DEAE-dextran

Acknowledgment

We would like to express our sincere gratitude to Dr. Jun Yoshida, Professor Emeritus of Nagoya University,for his stimulative advice. We also would like to thank Dr. George B. Butler, Professor Emeritus of University of Florida and the late Dr. Yasuo Kikuchi, Professor Emeritus of Nippon Bunri University.

References

[1] Saito,Y & Higuchi,Y & awakami,S & Yamashita,F & Hashida,M. (2009).Immunostimulatory Characteristics Induced by Linear Polyethyleneimine Plasmid DNA Complexes in Cultured Macrophages, *Human Gene Therapy*, **20**, 137-145.

[2] (1977).Dextran Fractions, Dextran Sulphate and DEAE-Dextran. Defined polymers for biological research:Uppsale, Sweden:Pharmacia Fine Chemicals.

[3] Anderson, E.C. & Master, R.C.& Mowat, G.N.(1971).Immune response of pigs of inactivated foot-and-mouth disease vaccines. Response to DEAE-Dextran and saponin adjuvated vaccines, *Res. Vet. Sci.*,**12**,351-357.

[4] Pagano, J.S.(1970).Biological activity of isolated viral nucleic acids, *Prog. Med. Virol.*, **12**, 1-48.

[5] Mumper,R.J. & Wang,J. & Claspell,J.M. & Roll, A.P. (1995).Novel polymeric condensing carriers for gene delivery, *Proc. Int. Symp. Control. Rel. Bioact. Mater.*, **22**, 178?179.

[6] Tovell, D.R. & Colter,J.S.(1967).Observations on the assay of infectious viral ribonucleic acid:effects of DMSO and DEAE-dextran, *Virology*.**32**, 84-92.

[7] Ryser, H.J.P. (1967).A membrane effect of basic polymers dependent on molecular size, *Nature*, **215**, 934-936.

[8] Ryser, H.J.P. (1968).Uptake of protein by mammalian cells: an underdeveloped area, *Science*, **159**, 390-396.

[9] Strauss, M. (1978).The incorporation of homologous and heterologous hypoxanthine-guanine phosphoribosyltransferase into mutant cells, *Biochim. Biophys. Acta*,**538**,11-22.

[10] Fox, R.M.& Mynderse, J.F & Goulian, M. (1977).Incorporation of deoxynucleotides into DNA by diethylaminoethyldextran-treated lymphocytes, *Biochemistry*,**16**,4470-4477.

[11] Borenfreund, E., et al.(1977).Diethylaminoethyl-Dextran and uptake of nucleic acids by mammalian cells, *J. Nat. Cancer Inst.*,**51**,1391-1392.

[12] Pagano, J.S.(1970).Biological activity of isolated viral nucleic acids, *Prog. Med. Virol.*,**12**,1-48.

[13] Pagano, J.S. & McCutchan, J.H.(1969).Enhancement of viral infectivity with DEAE-Dextran: Application to development of vaccines, *Prog. Immunobiol. Standard*,**3**,152-158.

[14] Rossi, C.R. & Kiesel, G.K.(1978).Bovine respiratory syncytial virus infection of bovine embryonic lung cultures: enhancement of infectivity with diethylaminoethyl-dextran and virus-infected cells, *Arch. Virol.*,**56**,227-236.

[15] Sasaki, K.& Furukawa, T. & Potkin, S.A.(1981).Enhancement of infectivity of cell-free variclla zoster virus with diethylaminoethyl dextran, *Proc. Soc. Exp. Biol. Med. (USA)*,**166**,281-286.

[16] Booth, J.C. (1977).Enhancement of diethylaminoethyl-dextran of the plaque-forming activity of foot-and-mouth disease virus-antibody complexes in pig kidney 1B-RS-2 cells, *Arch. Virol.*,**55**,251-261.

[17] Wallis, C. & Meiniek, J.L. (1968).Mechanism of enhancement of virus plaques by cationic polymers, *J. Virol.*,**2**,267-274.

[18] Ebbesen, P. (1974).Influence of DEAE-Dextran, polybrene. Dextran and Dextran Sulphate on spontaneous leukemia development in AKR mice and virus induced leukemia in Balb/C mice, *Brit. J. Cancer*,**30**,68-72.

[19] Thorling, E.B.& Larsen, B. & Nielsen, H. (1971).Inhibitory effect of DEAE-Dextran on tumour growth, *Acta Path. Microbiol. Scand. Section A*,**79**,81-90.

[20] Larsen, B.& Olsen,K. (1968).Inhibitory effect of polycations on the transplantability of mouse leukaemia reversed by heparin, *Eur. J. Cncer*,**4**,157-162.

[21] Larsen, B.& Olsen,K. (1969).Inhibitory effect of DEAE-Dextran on tumour growth. Action of dextran sulphate after in vitro incubation, *Acta Pathol. Microbiol. Scand.*,**75**,229-236.

[22] Clark, J.M. & Hirtenstein, M.D. (1981).High yield culture of human fibroblasts on microcarriers: a first step in production of fibroblast-derived interferon (human beta interferon), *J. Interferon Res.*,**1**,391-400.

[23] Pitha, P.M. & Carter, W.A. (1971).DEAE-Dextran: Poly IC complex. Physical properties and interferon, *Virology*,45,777-781.

[24] Trapman, J. (1979).A systematic study of interferon production by mouse L-929 cells induced with poly(I).Poly(C) and DEAE-Dextran, *FEBS Lett.*,**98**,107-110.

[25] Dianzani, F.& Gagnoni, S.& Cantagalli,P. (1970).Studies on the potentiating effect of DEAE-dextran on interferon production induced by double stranded synthetic ribonucleotides, *Ann. N.Y. Acad. Sci.*,**173**,727-735.

[26] Tilles, J.G. (1970).Diethylaminoethyl-dextran enhancement of interferon induction by a complex polynucleotide, *Proc. Soc. Exp. Biol. Med.*,**133**,1334-1341.

[27] Bausek, G.H.& Merigan, T. C. (1969).Cell interaction with a synthetic polynucleotide and interferon production in vitro, *Virology*,**39**,491-498.

[28] Houston, W.E., et al. (1975).Adjuvant effects of diethylaminoethyl-dextran, *Infect. Immun.*,**13**,1559-1562.

[29] Wittmann, G.& Dietzschold, B. & Bauer, K. (1975).Some investigations on the adjuvant mechanism of DEAE-Dextran, *Arch. Virol.*,**47**,225-235.

[30] Wittmann, G. (1970).The use of diethylaminoethyl-dextran (DEAE-D) as adjuvant for immunization of guinea pigs with inactivated foot-and-mouth disease (FMD) virus (Article in German.) , *Z. Bakt.Parasit. Infektionskr. Hyg., Abt. 1. Orig.*,**213**,1-8.

[31] Koch, G. & Bishop,J.M.(1968).The effect of polycations on the interaction of viral RNA with mammalian cells: studies on the infectivity of single- and double-stranded poliovirus RNA, *Virology*.**35**,9-17.

[32] Wentzky, P. & Koch,G.(1971).Influence of Polycations on the Interaction Between Poliovirus Multistranded Ribonucleic Acid and HeLa Cells, *J. Virol.* .**8**,35?40.

[33] Ohtani,K. & Nakamura,M. & Saito,S. & Nagata,K. & Sugamura,K. & Hinuma,Y.(1971).Electroporation: application to human lympboid cell lines for stable introduction of a transactivator gene of human T-cell leukemia virus type I, *Nucl.Acids Res.*.**17**,1589-1604.

[34] Fire,A. & Xu,S. & Montogomery,M.K. & Driver,S.A. & Mello,C.C.(1998).Potent and specific genetic interference by double-stranded RNA in Caenorhabditis elegans, *Nature*.**391**,806-810.

[35] Tsuboi, M. & Matsuo, K. & Ts'o, P. O. (1966).Interaction of poly-L-lysine and nucleic acids, *J. Mol. Biol.*.**15**, 256 ?267.

[36] Murata, J. & Ohya, Y. & Ouchi, T. (1966).Possibility of application of quaternary chitosan having pendant galactose residues as gene delivery tool, *Carbohydr. Polym.*,**29**,69-74.

[37] Sato, T. (1966).Carbohydrate Polymer for Gene Delivery, *Kobunshi*,**51**,837-840.

[38] McCutchan, JH. & Pagano, JS. (1968).Enhancement of the infectivity of simian virus 40 deoxyribonucleic acid with diethylaminoethyl-dextran, *J. Nat. Cancer Inst.*,**41**,351-358.

[39] Warden, D. ,Thorne. H.V. (1969).Influence of diethylaminoethyl-dextran on uptake and degradation of polyoma virus deoxyribonucleic acid by mouse embryo cells, *J Virol.*,**4**,380-387.

[40] Constantin, T. & Vendrely, C. (1969).Effect of DEAE-dextran on the incorporation of tritiated DNA by cultured rat cells, *CR. Soc. Bio.*,**4**,380-387.

[41] Mack, K. D.& Wei, R.& Elbagarri, A.& Abbey, N. & McGrath, S. D. (1998).A novel method for DEAE-dextran mediated transfection of adherent primary cultured human macrophages, *Immunol.methods*,**211**,79-86.

[42] Onishi, Y.& Kikuchi, Y. (2003).Study of the Complex between DNA and DEAE-Dextran, *Kobunshi Ronbunshu*,**60**,359-364.

[43] Onishi, Y.& Kikuchi, Y. (2004).Study of the Complex between RNA and DEAE-Dextran, *Kobunshi Ronbunshu*,**61**,139-143.

[44] Onishi, Y. & Maruno, S. & Kamiya, S. & Hokkoku, S. & Hasegawa, M. (1978).Preparation and characteristics of dextran-methyl methacrylate graft copolymer, *Polymer*,**19**,1325-1328.

[45] Onishi, Y. (1987).Cationic graft-copolymer, *U.S. Patent 4816540.*

[46] Onishi, Y. & Eshita, Y. & Murashita, A. & Mizuno, M. & Yoshida, J. (2005).Synthesis and Characterization of 2-diethylaminoethyl(DEAE)-dextran-MMA Graft Copolymer for Non-Viral Gene Delivery Vector, *J. Appl. Polym. Sci.*,**98**,9-14.

[47] Higashihara, J. & Onishi, Y. & Mizuno, M. & Yoshida, J. & Tamori, N. & Dieng, H. & Kato, K. & Okada, T. & Eshita, Y. (2005).Transfection of foreign genes into culture cells using novel DEAE-dextran copolymer as a non-viral gene carrier, *the 55th annual meeting of southern region, the Japan Society of Medical Entomology and Zoology; Miyazaki Prefecture Japan*,**55**,15.

[48] Onishi, Y. & Eshita, Y. & Murashita, A. & Mizuno, M. & Yoshida, J. (2006).2-Diethylaminoethyl(DEAE)-Dextran-MMA Graft Copolymer for Non-Viral Gene Delivery, *Bulletin of the Research Center of Environmental Science and Technology, Nippon Bunri University*,**5**,8-13.

[49] Onishi, Y. & Eshita, Y. & Murashita, A. & Mizuno, M. & Yoshida, J. (2007).Characteristics of 2-diethylaminoethyl(DEAE)-dextran-MMA graft copolymer as a non-viral gene carrier, *Nanomedicine: Nanotechnology, Biology and Medicine*,**3**,184-191.

[50] Onishi, Y. & Eshita, Y. & Murashita, A. & Mizuno, M. & Yoshida, J. (2008).A Novel Vector of 2-Diethylaminoethyl(DEAE)-Dextran-MMA Graft Copolymer for Non-Viral Gene Delivery, *The Journalof Gene Medicine*,**10**,430.

[51] Eshita, Y. & Higashihara, J. & Onishi, M. & Mizuno, M. & Yoshida, J.& Takasaki,T. & Kubota, N. & Onishi, Y. (2009).Mechanism of Introducing ExogenousGenes into Cultured Cells Using DEAE-Dextran- MMA Graft Copolymer as Non-Viral Gene Carrier, *Molecules*,**14**,2669-2683.

[52] Price, C. & Woods, D. A. (1973).A method for studying micellar aggregates in block and graft copolymers, *Euro Polymer Sci.*,**9**,827-830.

[53] Minagawa,K.& Matsuzawa, Y. & Yoshikawa, K. & Matsumoto, M. & Doi,M. (2002).Direct observation of the biphasic conformational change of DNA induced by cationic polymers, *FEBS Letters*,**295**,67-69.

[54] Yoshikawa, Y. & Emi, N. & Kanbe, T. & Yoshikawa, K.& Saito, H. (1996).Folding and aggregation of DNA chains induced by complexation with lipospermine: formation of a nucleosome-like structure and network assembly, *FEBS Letters*,**396**,71-76.

[55] Yamasaki,Y. & Yoshikawa, K. (1999).Collapsing Transition of Single DNA chains and Their Morphorogical Variations, *Kobunshi Ronbunshu*,**56**,772-785.

[56] Yoshikawa, Y. & Tsumoto, K. & Yoshikawa, K. (2002).Switching of Higher-Order Structure of DNA and Gene Expression, *Seibutsu Butsuri*,**42**,179-184.

[57] Behr,J. P. & Demeneix, B. & Loeffler,J. P. & Perez-Mutul. (1989).Efficient Gene Transfer into Mammalian Primary Endocrine Cells with Lipopolyamine-Coated DNA, *J. Proc Natl Acad Sci.*,**86**,6982-6986.

[58] Gao, X. & Huang, L. (1991).A novel cationic liposome reagent for efficient. transfection of mammalian cells, *Biochem Biophys Res Commun*,**179**,280-285.

[59] Kojima, H. & Ohishi, N. & Takamori, M.& Yagi,K. (1995).Cationic Multilamellar Liposome-Mediated Gene Transfer into Primary Myoblasts, *Biochem Biophys Res Commun*,**207**,8-12.

[60] Farber, F. E. & Melnick, J. L.& Butel,J. S. (1975).Optimal conditions for uptake of exogenous DNA by chinese hamster lung cells deficient in hypoxanthine-guanine phosphoribosyltransferase, *Biochim. Biophys. Acta.*,**390**,298-311.

[61] Holter, W. & Fordis, C. M. & Howard, B. H.(1989).Efficient gene transfer by sequential treatment of mammalian cells with DEAE-dextran and deoxyribonucleic acid, *Exp. Cell Res.*,**184**,546-551.

[62] Haensler, J. & Szoka, F. C.(1993).Polyamidoamine cascade polymers mediate efficient transfection of cells in culture, *Bioconjugate Chem.*,**4**,372 -379.

[63] Michaels, A S.(1965).Polyelectrolyte Complexes, *Ind. Eng. Chem.*,**57**,32- 40.

[64] Tanaka, S. & Shibayama, M. & Sasabe R. & Kawaguchi,H.(2003).Preparation and structure characterization of hairy nanoparticles consisting of hydrophobic core and thermosensitive hairs, *Polymer*,**44**,495-501.

[65] Horie, K. & Yamada, S. & Machida, S. & Takahashi, S. & Isono, Y. & Kawaguchi,H.(2003).Dansyl fluorescence and local structure of dansyl-labeled core-shell and core-hair type microspheres in solution, *Macromol. Chem. Phys.*,**204**,131-138.

[66] Onishi, Y. & Kamiya,S. & Nishioka, K.(1992).Study of Dextran Matrix Copolymer for Medical Use, *J.Jpn. C.L.*,**34**,289-298.

[67] Onishi, Y. & Maruno, S. & Hokkoku, S. (1979).Graft Copolymerization of Methyl Methacrylate onto Dextran and Some Properties of Copolymer, *Kobunshi Ronbunshu*,**36**,535-541.

[68] Graham, F. L. & van der Eb, A. J. (1973).A new technique for the assay of infectivity of human adenovirus 5 DNA, *Virology*,**52**,456-467.

[69] McCormik, C.L.& Park, L.S. (1981).Water-soluble copolymers. III. Dextran-g-poly(acrylamides) control of grafting sites and molecular weight by Ce(IV)-induced initiation in homogeneous solutions , *J.Polym.Sci.Polym.Chem.Ed.*,**19**,2229-2241.

[70] Onishi, Y. & Butler,G.B. & Hogen-Esch,T.E. (2004).1,2-propanediol-cellulose-acrylamide graft copolymers, *J. Appl. Polym. Sci.*,**92**,3022-3029.

[71] Schreier J B. (1969).Modification of deoxyribonuclease test medium for rapid identification of Serratia marcescens, *Am. J. Clin. Pathol.*,**51**,711-716.

[72] Ogren, P. J. & Henry, I. & Fletcher, S. E. S & Kelly, I.(2003).Chemical Applications of a Programmable Image Acquisition System, *J. Chem. Educ.*,**80**,699-703.

[73] Maes, R. & Sedwick, W. & Vaheri, A. (1967).Interaction between DEAE-dextran and nucleic acids, *Biochim. Biophys. Acta.*,**134**,269-276.

[74] Pagano, J. S. & Mc Cutchan, J. H. & Vaheri, A. (1967).Factors influencing the enhancement of the infectivity of poliovirus ribonucleic acid by diethylaminoethyl-dextran, *J. Virol.*,**1**,891-897.

[75] Bloomfield, V. A. (1997).DNA condensation by multivalent cations, *Biopolymers*,**44**,269-282.

[76] Chaszczewska-Markowska, M. & Ugorski, M. & Langner, M. (2004).Plasmid Condensation Induced by Cationic Compounds: Hydrophilic Polylysine and Amphiphilic Cationic Lipid, *Cell.Mol.Biol.Lett.*,**9**,3-13.

[77] Wedlock , D.N. & Keen, D.L. & McCarthy, A.R. & Andersen, P. & Buddle, B.M. (2002). Effect of different adjuvants on the immune responses of Mycobacterium tuberculosis culture filtrate proteins , *Vet Immunol Immunopathol* , **86**, 79-88.

[78] Youmans, A.S. & Youmans, G.P. (1972). Effect of Polybasic Amines on the Immunogenicity of Mycobacterial Ribonucleic Acid, *Infect Immun.*, **6**, 798-804.

Reviewer: Dr. Naoji Kubota, Professor, Dept. of Pharmacology and Therapeutics, Chemistry Faculty of Medicine, Oita University, Oita 879-5593, Japan.

In: Polymer Research and Applications
Editors: Andrew J. Fusco and Henry W. Lewis

ISBN: 978-1-61209-029-0

Commentary

BIO AND POLYMERS: NEW POLYMER TECHNOLOGIES WITH WATER

G. E. Zaikov [*] ***and L. L. Madyuskina***
N.M. Emanuel Institute of Biochemical Physics,
Russian Academy of Sciences, Moscow, Russia

"Bio and Polymer. New Polymer Technologies with Water" was held on September, 28 – 30 2008 in Aachen (Germany). This biannual meeting was organized by Gesellschaft Deutscher Chemiker, GDCh (German Chemical Society) and Fachgruppe Makromolekulare Chemie (Division of "Macromolecular chemistry").

The meeting was supported by Deutsche Farschungsgemeinschaft (DFG), the Chemical Company BASF, Bayer Material Science, Cognis, Evonik Industrial, LanXess, Henkel, Wacker. The scientific committee included world well known scientists and organizers of science: H.-W. Engels (Leverkusen), H. Heckroth (Leverkusen), B. Kuppers (Aachen), K. Landfester (Ulm), M. Moller (Aachen), G. Oenbrink (Marl), J. Sandler (Ludwigshafen), R. Schoenfeld (Dusseldorf), C. Birkner (GDCh, Frankfurt am Main), C. Dorr (GDCh, Frankfurt am Main).

The conference focused the exchange of new results on the interface of classical polymer science with biological systems and the importance of developing new water-based technologies. This includes topics ranging from "polymers from biomass", "chemical and physico-chemical transformations in water" to the "development of novel bio-inspired macromolecular systems and bio-hybrids". With the common theme expressed by the sub-title "New Polymer Technologies with Water", the conference spanned topics starting from application oriented research in water based polymers, to engineering polymers from biomass, and up to highly specific and novel bio-functional polymers.

Limited resources of water, energy, and materials cause a growing need for new concepts in material development and its ecological integration. Water technology and resource-management is a central factor not only in environmental protection, in health care, and agriculture, but also in energy production and for the availability of biomass as a material source. Within Polymer Science, this requires an adoption of a 'biological view' on polymer

[*] E-mail: chembio@sky.chph.ras.ru

synthesis, application, and the fate of polymer materials after use. Here, a key to future advancement is the ability to deal with water and to take advantage of its unique properties. Due to its dielectric properties, amphoteric nature and ability to undergo hydrogen bonding, water is certainly the most powerful solvent we know. It imposes enthalpic and entropic forces on other molecules that form an essential base for the richness in structure formation found in nature and the molecular functions in living systems. Hence, very topical scientific and technological challenges open up upon mastering water-based self-organisation of polymers; responsiveness, switching water solubility for water insolubility, and hybridisation of synthetic components with biological systems.

The program comprised almost 50 lectures with plenary papers given by:

- Dr. Stefan Marcinowski, Board member of BASF SE, Ludwigshafen
- Prof. Buddy Ratner, Director University of Washington Engineered Biomaterials
- Prof. Jean Frechet, Professor of Chemical Engineering, University of California
- Prof. Rolf Muelhaupt, Director Institute of Macromolecular Chemistry, Freiburg University
- Prof. Bernhard Rieger, WACKER-Chair of Macromolecular Chemistry, Technical University Munich

Dr. H-W. Engels (Chairman of the GDCh-division), Dr. E.M. Schmachtenberg (Rector of the RWTH Aachen) and Mr. J.Linden, (Mayor of Aachen) took part in opening ceremony of conference.

The plenary lecture of Dr. S. Marcinowski had title "Renewable Raw Materials – a Novel Approach in Polymer". Prof. J.M. Frechet spoke in his plenary lecture about water-compatible polymer carriers for therapeutics and Prof. B.D. Ratner gave information about water and biomaterials (hydrogels, healing and non-fouling).

Prof. B. Rieger presented information about carbon dioxide as building block of novel polymer architectures and Prof. R. Muelhaupt spoke about biofunctional materials (preparation, properties and application).

Several sessions were included in scientific program. The session "Bio-inspired Polymers" included 11 oral presentations. Self-assembly of an aquaporin mimic, tailoring surface properties with polymer brushes, bioinspired block copolymers, hierarchically structured conjugated polymers via supramolecular self-assembly, natural polymeric composites with mechanical function, macromolecular oxidation catalysts based on miniemulsion polymerization and some other problems were discussed on this session.

Second session had title "Bio-medical Polymers" and included 11 lectures. Participants of the conference had discussed the next problems: functionalized nanoparticles and nanocapsules as markers and nanocarriers in biomedical applications, smart hydrogels for biomedical applications, injectable biodegradable hydrogels for protein and cell delivery, biohybrid hydrogels for regenerative therapies, biodegradable polymers for biomedical applications, design and function of DNA and protein nanoparticles.

The session "Polymer from Biomass" included 8 oral presentations. These presentations were devoted to the problems of production (synthesis) polymers from biomass particularly: sustainability assessment of polymers based on renewable resources; novel cellulose based materials and processing routes; biomass-based polyesters and polycarbonates for coating and engineering plastic applications; plant oils as renewable resources in polymer science.

The problems of bio-functional polymers were discussed on the 4th session. This session included 6 lectures. The speakers gave information about biofunctional dendritic architectures, biocompatible and bioactive polymers containing saccharide functionality, design and mechanisms of antimicrobial polymers, control of protein adsorption on functionalized electrospun fibers, microcapsules and nanoparticles for controlled delivery and repair, smart nanocarriers for bioseparation and responsive drug delivery systems.

The last session "Waterborne Polymers" included 8 presentations. There were discussed the next problems: synthesis of smart nano-hydrogels, influence of adsorbed polymers on keratin surface properties in an aqueous environment, pH- and temperature-sensitive microgels for stimuli controlled emulsion stabilisation, industrial application of mini-emulsion polymerization in the coating field, PLURONIC® block copolymers in balanced microemulsions.

About 100 participants from 11 countries (Germany, USA, The Netherlands, Switzerland, China, Poland, Sweden, Finland, Austria, Russia and India) took part in this conference.

The two poster sessions included about 40 posters which were discussed some particular tasks in the field of new polymer technologies with water.

The meeting demonstrated that biopolymers are very important part of polymer science for today because the price for oil is going up too much.

The next similar conference will be in 2010.

INDEX

A

B

C

D

E

F

I

J

K

L

M

N

O

P

Q

R

S

T

U

V

W

Y

Z